Home Pork Making

by A. W. Fulton

TABLE OF CONTENTS.

INTRODUCTION.

Pork making on the farm nearly a lost art--General merit of homemade pork--Acknowledgments.

CHAPTER I.

--PORK MAKING ON THE FARM.

Best time for killing--A home market for farm pork--Opportunities for profit--Farm census of live stock for a series of years.

CHAPTER II.

--FINISHING OFF HOGS FOR BACON.

Flesh forming rations--Corn as a fat producer--Just the quality of bacon wanted--Normandy Hogs.

CHAPTER III.

--SLAUGHTERING.

Methods employed--Necessary apparatus--Heating water for scalding.

CHAPTER IV.

--SCALDING AND SCRAPING.

Saving the bristles--Scalding tubs and vats--Temperature for scalding--"Singeing pigs"--Methods of Singeing.

CHAPTER V.

--DRESSING AND CUTTING.

Best time for dressing--Opening the carcass--Various useful appliances--Hints on dressing--How to cut up a hog.

CHAPTER VI.

--WHAT TO DO WITH THE OFFAL.

Portions classed as offal--Recipes and complete directions for utilizing the wholesome parts, aside from the principal pieces--Sausage, scrapple, jowls and head, brawn, head-cheese.

CHAPTER VII.

--THE FINE POINTS IN MAKING LARD.

Kettle and steam rendered--Time required in making--Storing.

CHAPTER VIII.

--PICKLING AND BARRELING.

A clean barrel one of the first considerations--The use of salt on pork strips--Pickling by covering with brine--Renewing pork brine.

CHAPTER IX.

--CARE OF HAMS AND SHOULDERS.

A first-class ham--A general cure for ham and shoulders--Pickling preparatory to smoking--Westphalian hams.

CHAPTER X.

--DRY SALTING BACON AND SIDES.

Proper proportion of salt to meat--Other preservatives--Applying the salt--Best distribution of the salt--Time required in curing--Pork for the south.

CHAPTER XI.

--SMOKING AND SMOKEHOUSES.

Treatment previous to smoking--Simple but effective smokehouses--Controlling the fire in smoke formation--Materials to produce best flavor--The choice of weather--Variety in smokehouses.

CHAPTER XII.

--KEEPING HAMS AND BACON.

The ideal meat house--Best temperature and surroundings--Precautions against skippers--To exclude the bugs entirely.

CHAPTER XIII.

--SIDE LIGHTS ON PORK MAKING.

Growth of the big packing houses--Average weight of live hogs--"Net to gross"--Relative weights of various portions of the carcass.

CHAPTER XIV.

--PACKING HOUSE CUTS OF PORK.

Descriptions of the leading cuts of meat known as the speculative commodities in the pork product--Mess pork, short ribs, shoulders and hams, English bacon, varieties of lard.

CHAPTER XV.

--MAGNITUDE OF THE SWINE INDUSTRY.

Importance of the foreign demand--Statistics of the trade--Receipts at leading points--Prices for a series of years--Co-operative curing houses in Denmark.

CHAPTER XVI.

--DISCOVERING THE MERITS OF ROAST PIG.

The immortal Charles Lamb on the art of roasting--An oriental luxury of luxuries.

CHAPTER XVII.

--RECIPES FOR COOKING AND SERVING PORK.

Success in the kitchen--Prize methods of best cooks--Unapproachable list of especially prepared recipes--Roasts, pork pie, cooking bacon, pork and beans, serving chops and cutlets, use of spare ribs, the New England boiled dinner, ham and sausage, etc.

INTRODUCTION.

Hog killing and pork making on the farm have become almost lost arts in these days of mammoth packing establishments which handle such enormous numbers of swine at all seasons of the year. Yet the progressive farmer of to-day should not only provide his own fresh and cured pork for family use, but also should be able to supply at remunerative prices such persons in his neighborhood as appreciate the excellence and general merit of country or "homemade" pork product. This is true, also, though naturally in a less degree, of the townsman who fattens one or two pigs on the family kitchen slops, adding sufficient grain ration to finish off the pork for autumn slaughter.

The only popular book of the kind ever published, "Home Pork Making" furnishes in a plain manner just such detailed information as is needed to enable the farmer, feeder, or country butcher to successfully and economically slaughter his own hogs and cure his own pork. All stages of the work are fully presented, so that even without experience or special equipment any intelligent person can readily follow the instructions. Hints are given about finishing off hogs for bacon, hams, etc. Then, beginning with proper methods of slaughtering, the various processes are clearly presented, including every needful detail from the scalding vat to the kitchen baking dish and dining-room table.

The various chapters treat successively of the following, among other branches of the art of pork making: Possibilities of profit in home curing and marketing pork; finishing off hogs for bacon; class of rations best adapted, flesh and fat forming foods; best methods of slaughtering hogs, with necessary adjuncts for this preliminary work; scalding and scraping; the construction of vats; dressing the carcass; cooling and cutting up the meat; best disposition of the offal; the making of sausage and scrapple; success in producing a fine quality of lard and the proper care of it.

Several chapters are devoted to putting down and curing the different cuts of meat in a variety of ways for many purposes. Here will be found the prized recipes and secret processes employed in making the popular pork specialties for which England, Virginia, Kentucky, New England and other sections are noted. Many of these points involve the old and well-guarded methods upon which more than one fortune has been made, as well as the newest and latest ideas for curing pork and utilizing its products. Among these the subject of pickling and barreling is thoroughly treated, renewing pork brine; care of barrels, etc. The proper curing of hams and shoulders receives minute attention, and so with the work of dry salting bacon and sides. A chapter devoted to smoking and smokehouses affords all necessary light on this important subject, including a number of helpful illustrations; success in keeping bacon and hams is fully described, together with many other features of the work of home curing. The concluding portion of the book affords many interesting details relating to the various cuts of meat in the big packing houses, magnitude of the swine industry and figures covering the importance of our home and foreign trade in pork and pork product.

In completing this preface, descriptive of the various features of the book, the editor wishes to give credit to our friends who have added to its value through various contributions and courtesies. A considerable part of the chapters giving practical directions for cutting and curing pork are the results of the actual experience of B. W. Jones of Virginia; we desire also to give due credit to contributions by P. H. Hartwell, Rufus B. Martin, Henry Stewart and many other practical farmers; to Hately Brothers, leading packers at Chicago; North Packing and Provision Co. of Boston, and to a host of intelligent women on American farms, who, through their practical experience in the art of cooking, have furnished us with many admirable recipes for preparing and serving pork.

CHAPTER I.

PORK MAKING ON THE FARM.

During the marvelous growth of the packing industry the past generation, methods of slaughtering and handling pork have undergone an entire revolution. In the days of our fathers, annual hog-killing time was as much an event in the family as the harvesting of grain. With the coming of good vigorous frosts and cold weather, reached in the Northern states usually in November, every farmer would kill one, two or more hogs for home consumption, and frequently a considerable number for distribution through regular market channels. Nowadays, however, the big pork packing establishments have brought things down to such a fine point, utilizing every part of the animal (or, as has been said, "working up everything but the pig's squeal"), that comparatively few hogs out of all the great number fattened are slaughtered and cut up on the farm.

Unquestionably there is room for considerable business of this character, and if properly conducted, with a thorough understanding, farmers can profitably convert some of their hogs into cured meats, lard, hams, bacon, sausage, etc., finding a good market at home and in villages and towns. Methods now in use are not greatly different from those followed years ago, although of course improvement is the order of the day, and some important changes have taken place, as will be seen in a study of our pages. A few fixtures and implements are necessary to properly cure and pack pork, but these may be simple, inexpensive and at the same time efficient. Such important portions of the work as the proper cutting of the throat, scalding, scraping, opening and cleaning the hog should be undertaken by someone not altogether a novice. And there is no reason why every farmer should not advantageously slaughter one or more hogs each year, supplying the family with the winter's requirements and have something left over to sell.

THE POSSIBILITIES OF PROFIT

in the intelligent curing and selling of homemade pork are suggested by the far too general custom of farmers buying their pork supplies at the stores. This custom is increasing, to say nothing of the very large number of townspeople who would be willing to buy home cured pork were it properly offered them. Probably it is not practicable that every farmer should butcher his own swine, but in nearly every neighborhood one or two farmers could do this and make good profits. The first to do so, the first to be known as having home cured pork to sell, and the first to make a reputation on it, will be the one to secure the most profit.

In the farm census of live stock, hogs are given a very important place.

According to the United States census of 1890 there were on farms in this country 57,409,583 hogs. Returns covering later years place the farm census of hogs, according to compilations of American Agriculturist and Orange Judd Farmer, recognized authorities, at 47,061,000 in 1895, 46,302,000 in 1896, and 48,934,000 in 1899. According to these authorities the average farm value of all hogs in 1899 was $4.19 per head. The government report placed the average farm price in 1894 at $5.98, in '93, $6.41, and in 1892, $4.60.

A TRAVELING PIGPEN.

It is often desirable to change the location of a pigpen, especially where a single pig is kept. It may be placed in the garden at the time when there are waste vegetables to be disposed of, or it may be penned in a grass lot. A portable pen, with an open yard attached, is seen in the accompanying illustrations. Figure 1 presents the pen, the engraving showing it so clearly that no description is needed. The yard, seen in Fig. 2, is placed with the open space next to the door of the pen, so that the pig can go in and out freely. The yard is attached to the pen by hooks and staples, and both of them are provided with handles, by which they can be lifted and carried from place to place. Both the yard and pen should be floored, to prevent the pig from tearing up the ground. The floors should be raised a few inches from the ground, that they may be kept dry and made durable.

CHAPTER II.

FINISHING OFF HOGS FOR BACON.

The general subject of feeding and fattening hogs it is not necessary here to discuss. It will suffice to point out the advisability of using such rations as will finish off the swine in a manner best fitted to produce a good bacon hog. An important point is to feed a proper proportion of flesh-forming ration rather than one which will serve to develop fat at the expense of lean. The proper proportion of these will best subserve the interest of the farmer, whether he is finishing off swine for family use or for supplying the market with home cured bacon. A diet composed largely of protein (albuminoids) results in an increased proportion of lean meat in the carcass. On the other hand, a ration made up chiefly of feeds which are high in starchy elements, known as carbohydrates, yields very largely in fat (lard). A most comprehensive chart showing the relative values of various fodders and feeding stuffs has been prepared by Herbert Myrick, editor of American Agriculturist, and will afford a good many valuable hints to the farmer who wishes to feed his swine intelligently. This points out the fact that such feeds as oats, barley, cowpea hay, shorts, red clover hay and whole cottonseed are especially rich in flesh-forming properties.

Corn, which is rich in starch, is a great fat producer and should not be fed too freely in finishing off hogs for the best class of bacon. In addition to the important muscle-producing feeds noted above, there are others rich in protein, such as bran, skim milk, buttermilk, etc. While corn is naturally the standby of all swine growers, the rations for bacon purposes should include these muscle-producing feeds in order to bring the best results. If lean, juicy meat is desired, these muscle forming foods should be continued to the close. In order to get

JUST THE QUALITY OF BACON THAT IS WANTED,

feeders must so arrange the ration that it will contain a maximum of muscle and a minimum of fat. This gives the sweet flavor and streaked meat which is the secret of the popularity of the Irish and Danish bacon. Our American meats are as a rule heavy, rich in fat and in marked contrast with the light, mild, sweet flavored pork well streaked with lean, found so generally in the English market and cured primarily in Ireland and Denmark. What is wanted is a long, lean, smooth, bacon hog something after the Irish hog. Here is a hint for our American farmers.

England can justly boast of her hams and bacon, but for sweet, tender, lean pork the Normandy hogs probably have no superior in the world. They are fed largely on meat-producing food, as milk, peas, barley, rye and wheat bran.

They are not fed on corn meal alone. They are slaughtered at about six months. The bristles are burned off by laying the carcass on straw and setting it on fire. Though the carcasses come out black, they are scraped white and clean, and dressed perfectly while warm. It is believed that hogs thus dressed keep better and that the meat is sweeter.

SELF-CLOSING DOOR FOR PIGPEN.

Neither winter snows nor the spring and summer rains should be allowed to beat into a pigpen. But the difficulty is to have a door that will shut itself and can be opened by the animals whenever they desire. The engraving, Fig. 3, shows a door of this kind that can be applied to any pen, at least any to which a door can be affixed at all. It is hung on hooks and staples to the lintel of the doorway, and swinging either way allows the inmates of the pen to go out or in, as they please,--closing automatically. If the door is intended to fit closely, leather strips two inches wide should be nailed around the frame of the doorway, then as the door closes it presses tightly against these strips.

A HOG-FEEDING CONVENIENCE.

The usual hog's trough and the usual method of getting food into it are conducive to a perturbed state of mind on the part of the feeder, because the hog is accustomed to get bodily into the trough, where he is likely to receive a goodly portion of his breakfast or dinner upon the top of his head. The ordinary trough too, is difficult to clean out for a similar reason--the pig usually standing in it. The diagram shown herewith, Fig. 4 gives a suggestion for a trough that overcomes some of the difficulties mentioned, as it is easily accessible from the outside, both for pouring in food and for removing any dirt or litter that may be in it. The accompanying sketch so plainly shows the construction that detailed description does not appear to be necessary.

CHAPTER III.

SLAUGHTERING.

Whatever may be said as to the most humane modes of putting to death domestic animals intended for food, butchering with the knife, all things considered, is the best method to pursue with the hog. The hog should be bled thoroughly when it is killed. Butchering by which the heart is pierced or the main artery leading from it severed, does this in the most effectual way, ridding the matter of the largest percentage of blood, and leaving it in the best condition for curing and keeping well. The very best bacon cannot be made of meat that has not been thoroughly freed from blood, and this is a fact that should be well remembered. Expert butchers, who know how to seize and hold the hog and insert the knife at the proper place, are quickly through with the job, and often before the knife can be withdrawn from the incision, the blood will spurt out in a stream and insensibility and death will speedily ensue. It is easy, however, for a novice to make a botch of it; hence the importance that none but an expert be given a knife for this delicate operation.

There are some readily made devices by which one man at killing time may do as much as three or four, and with one helper a dozen hogs may be made into finished pork between breakfast and dinner, and without any excitement or worry or hard work. It is supposed that the hogs are in a pen or pens, where they may be easily roped by a noose around one hind leg. This being done, the animal is led to the door and guided into a box, having a slide door to shut it in. The bottom of the box is a hinged lid. As soon as the hog is safely in the box and shut in by sliding down the back door, and fastening it by a hook, the box is turned over, bringing the hog on his back. The bottom of the box is opened immediately and one man seizes a hind foot, to hold the animal, while the other sticks the hog in the usual manner. The box is turned and lifted from the hog, which, still held by the rope is moved to the dressing bench. All this may be done while the previous hog is being scalded and dressed, or the work may be so managed that as soon as one hog is hung and cleaned the next one is ready for the scalding.

NECESSARY AIDS.

Before the day for slaughter arrives, have everything ready for performing the work in the best manner. There may be a large boiler for scalding set in masonry with a fireplace underneath and a flue to carry off the smoke. If this is not available, a large hogshead may be utilized at the proper time. A long table, strong and immovable, should be fixed close to the boiler, on which the hogs are to be drawn after having been scalded, for scraping. On each side of this table scantlings should be laid in the form of an open flooring, and upon this the farmer and helpers may stand while at work, thus keeping their feet off the ground, out of the water and mud that would otherwise be disagreeable. An appreciated addition on a rainy day would be a substantial roof over this boiler and bench. This should be strong and large enough so that the hog after it is cleaned may be properly hung up. Hooks and gambrels are provided, knives are sharpened, a pile of dry wood is placed there, and everything that will be needed on the day of butchering is at hand.

HEATING WATER FOR SCALDING.

For heating scalding water and rendering lard, when one has no kettles or cauldrons ready to set in brick or stone, a simple method is to put down two forked stakes firmly, as shown in Fig. 5, lay in them a pole to support the kettles, and build a wood fire around them on the ground. A more elaborate arrangement is shown in Fig. 6, which serves not only to heat the water, but as a scalding tub as well. It is made of two-inch pine boards, six feet long and two feet wide, rounded at the ends. A heavy plate of sheet iron is nailed with wrought nails on the bottom and ends Let the iron project fully one inch on each side. The ends, being rounded, will prevent the fire from burning the woodwork. They also make it handier for dipping sheep, scalding hogs, or for taking out the boiled food. The box is set on two walls 18 inches high, and the rear end of the brickwork is built into a short chimney, affording ample draft.

CHAPTER IV.

SCALDING AND SCRAPING.

Next comes the scalding and dressing of the carcass. Lay the hog upon the table near the boiler and let the scalders who stand ready to handle it place it in the water heated nearly to a boiling point. The scalders keep the hog in motion by turning it about in the water, and occasionally they try the bristles to see if they will come away readily. As soon as satisfied on this point, the carcass is drawn from the boiler and placed upon the bench, where it is rapidly and thoroughly scraped. The bristles or hair that grow along the back of the animal are sometimes sold to brush makers, the remainder of the hair falling beside the table and gathered up for the manure heap. The carcass must not remain too long in the hot water, as this will set the hair. In this case it will not part from the skin, and must be scraped off with sharp knives. For this reason an experienced hand should attend to the scalding. The hair all off, the carcass is hung upon the hooks, head down, nicely scraped and washed with clean water preparatory to disemboweling.

SCALDING TUBS AND VATS.

Various devices are employed for scalding hogs, without lifting them by main force. For heavy hogs, one may use three strong poles, fastened at the top with a log chain, which supports a simple tackle, Fig. 7. A very good arrangement is shown in Fig. 8. A sled is made firm with driven stakes and covered with planks or boards. At the rear end the scalding cask is set in the ground, its upper edge on a level with the platform and inclined as much as it can be and hold sufficient water. A large, long hog is scalded one end at a time. The more the cask is inclined, the easier will be the lifting.

A modification of the above device is shown in Fig. 9. A lever is rigged like a well sweep, using a crotched stick for the post, and a strong pole for the sweep. The iron rod on which the sweep moves must be strong and stiff. A trace chain is attached to the upper end, and if the end of the chain has a ring instead of a hook, it will be quite convenient. In use, a table is improvised, unless a strong one for the purpose is at hand, and this is set near the barrel. A noose is made with the chain about the leg of the hog, and he is soused in, going entirely under water, lifted out when the bristles start easily, and laid upon the table, while another is made ready.

Figure 10 shows a more permanent arrangement. It is a trough of plank with a sheet iron bottom, which can be set over a temporary fireplace made in the ground. The vat may be six feet long, three feet wide and two and one-half feet deep, so as to be large enough for a good-sized hog. Three ropes are fastened on one side, for the purpose of rolling the hog over into the vat and rolling it out on the other side when it is scalded. A number of slanting crosspieces are fitted in, crossing each other, so as to form a hollow bed in which the carcass lies, with the ropes under it, by which it can be moved and drawn out. These crosspieces protect the sheet iron bottom and keep the carcass from resting upon it. A large, narrow fireplace is built up in the ground, with stoned sides, and the trough is set over it. A stovepipe is fitted at one end, and room is made at the front by which wood may be supplied to the fire to heat the water. A sloping table is fitted at one side for the purpose of rolling up the carcass, when too large to handle otherwise, by means of the rope previously mentioned. On the other side is a frame made of hollowed boards set on edge, upon which the hog is scraped and cleaned. The right temperature for scalding a hog is 180 degrees, and with a thermometer there need be no fear of overscalding or a failure from the lack of sufficient heat, while the water can be kept at the right temperature by regulating the fuel under the vat. If a spot of hair is obstinate, cover it with some of the removed hair and dip on hot water. Always pull out hair and bristles; shaving any off leaves unpleasant stubs in the skin.

SINGEING PIGS.

A few years ago, "singers" were general favorites with a certain class of trade wanting a light bacon pig, weighing about 170 lbs., the product being exported to England for bacon purposes. Packers frequently paid a small premium for light hogs suitable for this end, but more recently the demand is in other directions. The meat of singed hogs is considered by some to possess finer flavor than that of animals the hair of which has been removed by the ordinary process. Instead of being scalded and scraped in the ordinary manner, the singeing process consists in lowering the carcass into an iron or steel box by means of a heavy chain, the receptacle having been previously heated to an exceedingly high temperature. After remaining there a very few seconds the hog is removed and upon being placed in hot water the hair comes off instantly.

An old encyclopedia, published thirty years ago, in advocating the singeing process, has this to say: "The hog should be swealed (singed), and not scalded, as this method leaves the flesh firm and more solid. This is done by covering the hog lightly with straw, then set fire to it, renewing the fuel as it is burned away, taking care not to burn the skin. After sufficient singeing, the skin is scraped, but not washed. After cutting up, the flesh side of the cuts is rubbed with salt, which should be changed every four or five days. The flitches should also be transposed, the bottom ones at the top and the top ones at the bottom. Some use four ounces saltpetre and one pound coarse sugar or molasses for each hog. Six weeks is allowed for thus curing a hog weighing 240 lbs. The flitches before smoking are rubbed with bran or very fine sawdust and after smoking are often kept in clear, dry wood ashes or very dry sand."

CHAPTER V.

DRESSING AND CUTTING.

When the carcasses have lost the animal heat they are put away till the morrow, by which time, if the weather is fairly cold, the meat is stiff and firm and in a condition to cut out better than it does when taken in its soft and pliant state. If the weather is very cold, however, and there is danger that the meat will freeze hard before morning, haste is made to cut it up the same day, or else it is put into a basement or other warm room, or a large fire made near it to prevent it from freezing. Meat that is frozen will not take salt, or keep from spoiling if salted. Salting is one of the most important of the several processes in the art of curing good bacon, and the pork should be in just the right condition for taking or absorbing the salt. Moderately cold and damp weather is the best for this.

AS THE CARCASS IS DRESSED

it is lifted by a hook at the end of a swivel lever mounted on a post and swung around to a hanging bar, placed conveniently. This bar has sliding hooks made to receive the gambrel sticks, which have a hook permanently attached to each so that the carcass is quickly removed from the swivel lever to the slide hook on the bar. The upper edge of the bar is rounded and smoothed and greased to help the hooks to slide on it. This serves to hang all the hogs on the bar until they are cooled. If four persons are employed this work may be done very quickly, as they may divide the work between them; one hog is being scalded and cleaned while another is being dressed.

Divested of its coat, the carcass is washed off nicely with clean water before being disemboweled. For opening the hog, the operator needs a sharp butcher's knife, and should know how to use it with dexterity, so as not to cut the entrails. The entrails and paunch, or stomach, are first removed, care being taken not to cut any; then the liver, the "dead ears" removed from the heart, and the heart cut open to remove any clots of blood that it may contain. The windpipe is then slit open, and the whole together is hung upon the gambrel beside the hog or placed temporarily into a tub of water. The "stretcher," a small stick some sixteen inches long, is then placed across the bowels to hold the sides well open and admit the air to cool the carcass, and a chip or other small object is placed in the mouth to hold it open, and the interior parts of the hog about the shoulders and gullet are nicely washed to free them from stains of blood. The carcass is then left to hang upon the gallows in order to cool thoroughly before it is cut into pieces or put away for the night.

Where ten or twelve hogs are dressed every year, it will pay to have a suitable building arranged for the work. An excellent place may be made in the driveway between a double corncrib, or in a wagon shed or an annex to the barn where the feeding pen is placed. The building should have a stationary boiler in it, and such apparatus as has been suggested, and a windlass used to do the lifting.

HOG KILLING MADE EASY.

In the accompanying cut, Fig. 11, the hoister represents a homemade apparatus that has been in use many years and it has been a grand success. The frames, a, a, a, a, are of 2x4 inch scantling, 8 ft. in length; b, b, are 2x6 inch and 2 ft. long with a round notch in the center of the upper surface for a windlass, d, to turn in; c, c are 2x4 and 8 ft. long, or as long as desired, and are bolted to a, a. Ten inches beyond the windlass, d, is a 4x4 inch piece with arms bolted on the end to turn the windlass and draw up the carcass, which should be turned lengthwise of the hoister until it passes between c, c. The gambrel should be long enough to catch on each side when turned crosswise, thus relieving the windlass so that a second carcass may be hoisted. The peg, e, is to place in a hole of upright, a, to hold the windlass. Brace the frame in proportion to the load that is to be placed upon it. The longer it is made, the more hogs can be hung at the same time.

THE SAWBUCK SCAFFOLD.

Figure 12 shows a very cheap and convenient device for hanging either hogs or beeves. The device is in shape much like an old-fashioned "sawbuck," with the lower rounds between the legs omitted. The legs, of which there are two pairs, should be about ten feet long and set bracing, in the manner shown in the engraving. The two pairs of legs are held together by an inch iron rod, five or six feet in length, provided with threads at both ends. The whole is made secure by means of two pairs of nuts, which fasten the legs to the connecting iron rod. A straight and smooth wooden roller rests in the forks made by the crossing of the legs, and one end projects about sixteen inches. In this two augur holes are bored, in which levers may be inserted for turning the roller. The rope, by means of which the carcass is raised, passes over the rollers in such a way that in turning, by means of the levers, the animal is raised from the ground. When sufficiently elevated, the roller is fastened by one of the levers to the nearest leg.

PROPER SHAPE OF GAMBRELS.

Gambrels should be provided of different lengths, if the hogs vary much in size. That shown in Fig. 13 is a convenient shape. These should be of hickory or other tough wood for safety, and be so small as to require little gashing of the legs to receive them.

GALLOWS FOR DRESSED HOGS.

The accompanying device, Fig. 14, for hanging dressed hogs, consists of a stout, upright post, six or eight inches square and ten feet long, the lower three feet being set into the ground. Near the upper end are two mortises, each 2x4 inches, quite through the post, one above the other, as shown in the engraving, for the reception of the horizontal arms. The latter are six feet long and just large enough to fit closely into the mortises. They should be of white oak or hickory. At butchering time the dead hogs are hung on the scaffold by slipping the gambrels upon the horizontal crosspieces.

ADDITIONAL HINTS ON DRESSING.

Little use of the knife is required to loosen the entrails. The fingers, rightly used, will do most of the severing. Small, strong strings, cut in proper lengths, should be always at hand to quickly tie the severed ends of any small intestines cut or broken by chance. An expert will catch the entire offal in a large tin pan or wooden vessel, which is held between himself and the hog. Unskilled operators, and those opening very large hogs, need an assistant to hold this. The entrails and then the liver, heart, etc., being all removed, thoroughly rinse out any blood or filth that may have escaped inside. Removing the lard from the long intestines requires expertness that can be learned only by practice. The fingers do most of this cleaner, safer and better than a knife. A light feed the night before killing leaves the intestines less distended and less likely to be broken.

HOW TO CUT UP A HOG.

With a sharp ax and a sharp butcher's knife at hand, lay the hog on the chopping bench, side down. With the knife make a cut near the ear clear across the neck and down to the bone. With a dextrous stroke of the ax sever the head from the body. Lay the carcass on the back, a boy holding it upright and keeping the forelegs well apart. With the ax proceed to take out the chine or backbone. If it is desired to put as much of the hog into neat meat as possible, trim to the chine very close, taking out none of the skin or outside fat with it. Otherwise, the cutter need not be particular how much meat comes away with the bone. What does not go with the neat meat will be in the offal or sausage, and nothing will be lost. Lay the chine aside and with the knife finish separating the two divisions of the hog. Next, strip off with the hands the leaves or flakes of fat from the middle to the hams. Seize the hock of the ham with the left hand and with the knife in the other, proceed to round out the ham, giving it a neat, oval shape. Be very particular in shaping the ham. If it is spoiled in the first cutting, no subsequent trimming will put it into a form to exactly suit the fastidious public eye. Trim off the surplus lean and fat and projecting pieces of bone. Cut off the foot just above the hock joint. The piece when finished should have nearly the form of a regular oval, with its projecting handle or hock.

With the ax cut the shoulder from the middling, making the cut straight across near the elbow joint. Take off the end ribs or "spare bone" from the shoulder, trim the piece and cut off the foot. For home use, trim the shoulder, as well as the other pieces, very closely, taking off all of both lean and fat that can be spared. If care is taken to cut away the head near the ear, the shoulder will be at first about as wide as long, having a good deal of the neck attached. If the meat is intended for sale and the largest quantity of bacon is the primary object, let the piece remain so. But if it is preferred to have plenty of lard and sausage, cut a smart strip from off the neck side of the shoulder and make the piece assume the form of a parallelogram, with the hock attached to one end. Trim a slice of fat from the back for lard, take off the "short ribs," and, if preferred, remove the long ribs from the whole piece. The latter, however, is not often done by the farmers. Put the middling in nice shape by trimming it wherever needed, which, when finished, will be very much like a square in form, perhaps a little longer than broad, with a small circular piece cut out from the end next the ham.

The six pieces of neat meat are now ready for the salter. The head is next cut open longitudinally from side to side, separating the jowl from the top or "head," so-called. The jawbone of the jowl is cut at the angle or tip and the "swallow," which is the larynx or upper part of the windpipe, is taken out. The headpiece is next cut open vertically and the lobe of the brain is taken out, and the ears and nose are removed.

The bone of the chine is cut at several places for the convenience of the cook, and the task of the cutter is finished. Besides the six pieces of neat meat, there are the chine, souse, jowl, head, fat, sausage, two spare and two short ribs and various other small bits derived from each hog. A good cutter, with an assistant to carry away the pieces and help otherwise, can cut out from 50 to 60 hogs in a day.

CHAPTER VI.

WHAT TO DO WITH THE OFFAL.

Aside from the pieces of meat into which a hog is usually cut, there will be left as offal the chine or backbone, the jowl, the souse, the liver and lungs, pig's feet, two spareribs and two short ribs or griskins. Nearly every housekeeper knows what disposition to make of all this, yet too often these wholesome portions of the hog are not utilized to best advantage.

PORK SAUSAGE.

Sausage has formed a highly prized article of food for a good many hundred years. Formed primarily as now, by chopping the raw meat very fine, and adding salt and other flavoring materials, and often meal or bread crumbs, the favorite varieties of to-day might not be considered any improvement over the recipes of the ancient Romans were they to pass judgment on the same. History tells us that these early Italian sausages were made of fresh pork and bacon, chopped fine, with the addition of nuts, and flavored with cumin seed, pepper, bay leaves and various pot herbs. Italy and Germany are still celebrated for their bologna sausages and with many people these smoked varieties are highly prized.

Like pure lard, sausage is too often a scarce article in the market. Most city butchers mix a good deal of beef with the pork, before it is ground, and so have a sausage composed of two sorts of meat, which does not possess that agreeable, sweet, savory taste peculiar to nice fresh pork. The bits of lean, cut off when trimming the pieces of neat meat, the tenderloins, and slices of lean from the shoulders and hams, together with some fat, are first washed nicely, cleared of bone and scraps of skin, then put into the chopper, and ground fine. If a great deal of sausage is wanted, the neat meat is trimmed very close, so as to take all the lean that can be spared from the pieces. Sometimes whole shoulders are cut up and ground. The heads, too, or the fleshy part, make good sausage. Some housekeepers have the livers and "lights," or lungs, ground up and prepared for sausage, and they make a tolerable substitute. This preparation should be kept separate from the other, however, and be eaten while cold weather lasts, as it will not keep as long as the other kind.

After sausage is properly ground, add salt, sage, rosemary, and red or black pepper to suit the taste. The rosemary may be omitted, but sage is essential. All these articles should be made fine before mixing them with the meat. In order to determine accurately whether the sausage contains enough of these ingredients, cook a little and taste it.

If sausage is to be kept in jars, pack it away closely in them, as soon as it is ground and seasoned, and set the jars, securely closed, in a cool room. But it is much better to provide for smoking some of it, to keep through the spring and early summer. When the entrails are ready, stuff them full with the meat, after which the ends are tied and drawn together, and the sausage hung up in the smokehouse for smoking. This finishes the process of making pork sausage. Put up in this way, it deserves the name of sausage and it makes a dish good enough for any one. It is one of the luxuries of life which may be manufactured at home.

BOLOGNA SAUSAGE.

The popular theory is that these familiar sausages originated in the Italian city of that name, where the American visitor always stops for a bit of "the original." Many formulas are used in the preparation of bologna sausages, or rather many modifications of a general formula. Lean, fresh meat trimmings are employed and some add a small proportion of heart, all chopped very fine. While being chopped, spices and seasoning are added, with a sufficient quantity of salt. The meat employed is for the most part beef, to which is added some fresh or salted pork. When almost completed, add gradually a small quantity of potato flour and a little water. The mixture being of the proper consistency, stuff in beef casings, tie the ends together into rings of fair length and smoke thoroughly. This accomplished, boil until the sausages rise to the top, when they are ready for use. Some recipes provide for two parts of beef and one part of fat pork and the addition of a little ground coriander seed to the seasoning.

WESTPHALIAN SAUSAGES

are made in much the same manner as frankforts, chopped not quite so fine, and, after being cased, are smoked about a week.

FRANKFORT SAUSAGES.

Clean bits of pork, both fat and lean, are chopped fine and well moistened with cold water. These may be placed in either sheep or hog casings through the use of the homemade filler shown on another page.

SUABIAN SAUSAGES.

Chop very finely fat and lean meat until the mass becomes nearly a paste, applying a sprinkling of cold water during the operation. Suabian sausages are prepared by either smoking or boiling, and in the latter case may be considered sufficiently cooked when they rise to the surface of the water in which they are boiled.

ITALIAN PORK SAUSAGES.

The preparation of these requires considerable care, but the product is highly prized by many. For every nine pounds of raw pork add an equal amount of boiled salt pork and an equal amount of raw veal. Then add two pounds selected sardines with all bones previously removed. Chop together to a fine mass and then add five pounds raw fat pork previously cut into small cubes. For the seasoning take six ounces salt, four ounces ground pepper, eight ounces capers, eight ounces pistachio nuts peeled and boiled in wine. All of these ingredients being thoroughly mixed, add about one dozen pickled and boiled tongues cut into narrow strips. Place the sausage in beef casings of good size. In boiling, the sausages should be wrapped in a cloth with liberal windings of stout twine and allowed to cook about an hour. Then remove to a cool place about 24 hours.

TONGUE SAUSAGE.

To every pound of meat used add two pounds of tongues, which have previously been cut into small pieces, mixing thoroughly. These are to be placed in large casings and boiled for about an hour. The flavor of the product may be improved if the tongues are previously placed for a day in spiced brine. Pickled tongues are sometimes used, steeped first in cold water for several hours.

BLACK FOREST SAUSAGES.

This is an old formula followed extensively in years gone by in Germany. Very lean pork is chopped into a fine mass and for every ten pounds, three pounds of fat bacon are added, previously cut comparatively fine. This is properly salted and spiced and sometimes a sprinkling of blood is added to improve the color. Fill into large casings, place over the fire in a kettle of cold water and simmer without boiling for nearly an hour.

LIVER SAUSAGE.

The Germans prepare this by adding to every five pounds of fat and lean pork an equal quantity of ground rind and two and one-half pounds liver. Previously partly cook the rind and pork and chop fine, then add the raw liver well chopped and press through a coarse sieve. Mix all thoroughly with sufficient seasoning. As the raw liver will swell when placed in boiling water, these sausages should be filled into large skins, leaving say a quarter of the space for expansion. Boil nearly one hour, dry, then smoke four or five days.

ROYAL CAMBRIDGE SAUSAGES

are made by adding rice in the proportion of five pounds to every ten pounds of lean meat and six pounds of fat. Previously boil the rice about ten minutes, then add gradually to the meat while being chopped fine, not forgetting the seasoning. The rice may thus be used instead of bread, and it is claimed to aid in keeping the sausages fresh and sweet.

BRAIN SAUSAGES.

Free from all skin and wash thoroughly the brain of two calves. Add one pound of lean and one pound of fat pork previously chopped fine. Use as seasoning four or five raw grated onions, one ounce salt, one-half ounce ground pepper. Mix thoroughly, place in beef casings and boil about five minutes. Afterward hang in a cool place until ready for use.

TOMATO SAUSAGES.

Add one and one-half pounds pulp of choice ripe tomatoes to every seven pounds of sausage meat, using an addition of one pound of finely crushed crackers, the last named previously mixed with a quart of water and allowed to stand for some time before using. Add the mixture of tomato and cracker powder gradually to the meat while the latter is being chopped. Season well and cook thoroughly.

SPANISH SAUSAGE

is made by using one-third each leaf lard, lean and fat pork, first thoroughly boiling and chopping fine the meat. Add to this the leaf lard previously chopped moderately fine, mix well and add a little blood to improve the color and moisten the whole. This sausage is to be placed in large casings and tied in links eight to twelve inches long. In an old recipe for Spanish sausage seasoning it is made of seven pounds ground white pepper, six ounces ground nutmeg, eight ounces ground pimento or allspice and a sprinkling of bruised garlic.

ANOTHER SAUSAGE SEASONING.

To five pounds salt add two pounds best ground white pepper, three ounces ground mace, or an equal quantity of nutmeg, four ounces ground coriander seed, two ounces powdered cayenne pepper and mix thoroughly.

ADMIXTURE OF BREAD.

Very often concerns which manufacture sausage on a large scale add considerable quantities of bread. This increases the weight at low cost, thus cheapening the finished product, and is also said to aid in keeping qualities. While this is no doubt thoroughly wholesome, it is not in vogue by our most successful farmers who have long made a business of preparing home-cured sausage. Bread used for sausages should have the crust removed, should be well soaked in cold water for some time before required, then pressed to remove the surplus moisture, and added gradually to the pork while being chopped. Some sausage manufacturers add 10 to 15 per cent in weight of crushed crackers instead of bread to sausage made during hot weather. This is to render the product firm and incidentally to increase the weight through thoroughly mixing the cracker crumbs or powder with an equal weight or more of water before adding to the meat.

SAUSAGE IN CASES.

Many prefer to pack in sausage casings, either home prepared or purchased of a dealer in packers' supplies. Latest improved machines for rapidly filling the cases are admirably adapted to the work, and this can also be accomplished by a homemade device. Figure 15 shows a simple bench and lever arrangement to be used with the common sausage filler, which lightens the work so much that even a small boy can use it with ease, and any person can get up the whole apparatus at home with little or no expense. An inch thick pine board one foot wide and four and one-fourth feet long is fitted with four legs, two and one-half feet long, notched into its edges, with the feet spread outward to give firmness. Two oak standards eighteen inches high are set thirty-four inches apart, with a slot down the middle of each, for the admission of an oak lever eight feet long. The left upright has three or four holes, one above another, for the lever pin, as shown in the engraving. The tin filler is set into the bench nearer the left upright and projects below for receiving the skins. Above the filler is a follower fitting closely into it, and its top working very loosely in the lever to allow full play as it moves up and down. The engraving shows the parts and mode of working.

[Illustration: FIG. 15. HOMEMADE SAUSAGE FILLER.]

PHILADELPHIA SCRAPPLE.

This is highly prized in some parts of the country, affording a breakfast dish of great relish. A leading Philadelphia manufacturer has furnished us with the following recipe: To make 200 lbs. of scrapple, take about 80 lbs. of good clean pork heads, remove the eyes, brains, snout, etc. Put in about 20 gals. of water and cook until it is thoroughly done. Then take out, separate the bones and chop the meat fine. Take about 15 gals, of the liquor left after boiling the heads, and if the water has boiled down to a quantity less than 15 gals., make up its bulk with hot water; if more than 15 gals. remain, take some of the water out, but be sure to keep some of the good fat liquor. Put this quantity of the liquor into a kettle, add the chopped meat, together with 10 oz. pure white pepper, 8 oz. sweet marjoram, 2 lbs. fine salt. Stir well until the liquor comes to a good boil. Have ready for use at this time 25 lbs. good Indian meal and 7 lbs. buckwheat flour. As soon as the liquor begins to boil add the meal and flour, the two being previously mixed dry. Be careful to put the meal in a little at a time, scattering it well and stirring briskly, that it may not burn to the kettle. Cook until well done, then place in pans to cool. The pans should be well greased, also the dipper used, to prevent the scrapple sticking to the utensils. When cold, the scrapple is cut into slices and fried in the ordinary manner as sausage. Serve hot.

SOUSE.

After being carefully cleaned and soaked in cold water, the feet, ears, nose and sometimes portions of the head may be boiled, thoroughly boned, and pressed into bowls or other vessels for cake souse. But frequently these pieces, instead of being boned, are placed whole in a vessel and covered with a vinegar, and afterwards taken a little at a time, as wanted, and fried.

JOWLS AND HEAD.

If not made into souse or sausage, these may be boiled unsmoked, with turnips, peas or beans; or smoked and cooked with cabbage or salad. The liver and accompanying parts, if not converted into sausage, may be otherwise utilized.

THE SPARERIBS AND SHORT BONES

may be cooked in meat pies with a crust, the same as chicken, or they may be fried or boiled. The large end of the chine makes a good piece for baking. The whole chine may be smoked and will keep a long time.

CRACKNELS.

This is the portion of the fat meat which is left after the lard is cooked, and is used by many as an appetizing food. The cracknels may be pressed and thus much more lard secured. This latter, however, should be used before the best lard put away in tubs. After being pressed the cracknels are worked into a dough with corn meal and together made into cracknel bread.

BRAWN

is comparatively little used in this country, though formerly a highly relished dish in Europe, where it was often prepared from the flesh of the wild boar. An ancient recipe is as follows: "The bones being taken out of the flitches (sides) or other parts, the flesh is sprinkled with salt and laid on a tray, that the blood may drain off, after which it is salted a little and rolled up as hard as possible. The length of the collar of brawn should be as much as one side of the boar will permit; so that when rolled up the piece may be nine or ten inches in diameter. After being thus rolled up, it is boiled in a copper or large kettle, till it is so tender that you may run a stiff straw through it. Then it is set aside till it is thoroughly cold, put into a pickle composed of water, salt, and wheat-bran, in the proportion of two handfuls of each of the latter to every gallon of water, which, after being well boiled together, is strained off as clear as possible from the bran, and, when quite cold, the brawn is put into it."

HEAD CHEESE.

This article is made usually of pork, or rather from the meat off the pig's head, skins, and coarse trimmings. After having been well boiled, the meat is cut into pieces, seasoned well with sage, salt, and pepper, and pressed a little, so as to drive out the extra fat and water. Some add the meat from a beef head to make it lean. Others add portions of heart and liver, heating all in a big pan or other vessel, and then running through a sausage mill while hot.

BLOOD PUDDINGS

are usually made from the hog's blood with chopped pork, and seasoned, then put in casings and cooked. Some make them of beef's blood, adding a little milk; but the former is the better, as it is thought to be the richer.

SPICED PUDDINGS.

These are made somewhat like head-cheese, and often prepared by the German dealers, some of whom make large quantities. They are also made of the meat from the pig's chops or cheeks, etc., well spiced and boiled. Some smoke them.

CHAPTER VII.

THE FINE POINTS IN MAKING LARD.

Pure lard should contain less than one per cent of water and foreign matter. It is the fat of swine, separated from the animal tissue by the process of rendering. The choicest lard is made from the whole "leaf." Lard is also made by the big packers from the residue after rendering the leaf and expressing a "neutral" lard, which is used in the manufacture of oleomargarine. A good quality of lard is made from back-fat and leaf rendered together. Fat from the head and intestines goes to make the cheaper grades. Lard may be either "kettle" or "steam rendered," the kettle process being usually employed for the choicer fat parts of the animal, while head and intestinal fat furnish the so-called "steam lard." Steam lard, however, is sometimes made from the leaf. On the other hand, other parts than the leaf are often kettle rendered. Kettle rendered lard usually has a fragrant cooked odor and a slight color, while steam lard often has a strong animal odor.

TO REFINE LARD,

a large iron pot is set over a slow fire of coals, a small quantity of water is put into the bottom of the pot, and this is then filled to the brim with the fat, after it has first been cut into small pieces and nicely washed, to free it from blood and other impurities. If necessary to keep out soot, ashes, etc., loose covers or lids are placed over the vessels, and the contents are made to simmer slowly for several hours. This work requires a careful and experienced hand to superintend it. Everything should be thoroughly clean, and the attendant must possess patience and a practical knowledge of the work. It will not do to hurry the cooking. A slow boil or simmer is the proper way. The contents are occasionally stirred as the cooking proceeds, to prevent burning. The cooking is continued until the liquid ceases to bubble and becomes clear. So long as there is any milky or cloudy appearance about the fat, it contains water, and in this condition will not keep well in summer--a matter of importance to the country housekeeper.

It requires six to eight hours constant cooking to properly refine a kettle or pot of fat. The time will depend, of course, somewhat upon the size of the vessel containing it and the thickness of the fat, and also upon the attention bestowed upon it by the cook. By close watching, so as to keep the fire just right all the time, it will cook in a shorter period, and vice versa. When the liquid appears clear the pots are set aside for the lard to cool a little before putting it into the vessels in which it is to be kept. The cracknels are first dipped from the pots and put into colanders, to allow the lard to drip from them. Some press the cracknels, and thus get a good deal more lard. As the liquid fat is dipped from the pots it is carefully strained through fine colanders or wire sieves. This is done to rid it of any bits of cracknel, etc., that may remain in the lard. Some country people when cooking lard add a few sprigs of rosemary or thyme, to impart a pleasant flavor to it. A slight taste of these herbs is not objectionable. Nothing else whatever is put into the lard as it is cooked, and if thoroughly done, nothing else is needed. A little salt is sometimes added, to make it firmer and keep it better in summer, but the benefit, if any, is slight, and too much salt is objectionable.

LEAF LARD.

In making lard, all the leaf or flake fat, the two leaves of almost solid fat that grow just above the hams on either side about the kidneys, and the choice pieces of fat meat cut off in trimming the pork should be tried or rendered first and separate from the remainder. This fat is the best and makes what is called the leaf lard. It may be put in the bottom of the cans, for use in summer, or else into separate jars or cans, and set away in a cool place. The entrail fat and bits of fat meat are cooked last and put on top of the other, or into separate vessels, to be used during cool weather. This lard is never as good as the other, and will not keep sweet as long; hence the pains taken by careful housewives to keep the two sorts apart. It must be admitted, however, that many persons, when refining lard for market, do not make any distinction, but lump all together, both in cooking and afterward. But for pure, honest "leaf" lard not a bit of entrail fat should be mixed with the flakes.

A PARTICULARLY IMPORTANT POINT

in making lard is to take plenty of time. The cooking must not be hurried in the least. It requires time to thoroughly dry out all the water, and the keeping quality of the lard depends largely upon this. A slow fire of coals only should be placed under the kettle, and great care exercised that no spark snaps into it, to set fire to the hot oil. It is well to have at hand some close-fitting covers, to be put immediately over the kettle, closing it tightly in case the oil should take fire. The mere exclusion of air will put out the fire at once. Cook slowly in order not to burn any of the fat in the least, as that will impart a very unpleasant flavor to the lard. The attendants should stir well with a long ladle or wooden stick during the whole time of cooking. It requires several hours to thoroughly cook a vessel of lard, when the cracknels will eventually rise to the top.

A cool, dry room, such as a basement, is the best place for keeping lard.

Large stone jars are perhaps the best vessels to keep it in, but tins are cheaper, and wooden casks, made of oak, are very good. Any pine wood, cedar or cypress will impart a taste of the wood. The vessels must be kept closed, to exclude litter, and care should be observed to prevent ants, mice, etc., from getting to the lard. A secret in keeping lard firm and good in hot weather is first to cook it well, and then set it in a cool, dry cellar, where the temperature remains fairly uniform throughout the year. Cover the vessels after they are set away in the cellar with closely fitting tops over a layer of oiled paper.

CHAPTER VIII.

PICKLING AND BARRELING.

For salt pork, one of the first considerations is a clean barrel, which can be used over and over again after yearly renovation. A good way to clean the barrel is to place about ten gallons of water and a peck of clean wood ashes in the barrel, then throw in well-heated irons, enough to boil the water, cover closely, and by adding a hot iron occasionally, keep the mixture boiling a couple of hours. Pour out, wash thoroughly with fresh water, and it will be as sweet as a new barrel. Next cover the bottom of the barrel with coarse salt, cut the pork into strips about six inches wide, stand edgewise in the barrel, with the skin next the outside, until the bottom is covered. Cover with a thick coat of salt, so as to hide the pork entirely. Repeat in the same manner until the barrel is full, or the pork all in, covering the top thickly with another layer of salt. Let stand three or four days, then put on a heavy flat stone and sufficient cold water to cover the pork. After the water is on, sprinkle one pound best black pepper over all. An inch of salt in the bottom and between each layer and an inch and a half on top will be sufficient to keep the pork without making brine.

When it is desired to pickle pork by pouring brine over the filled barrel, the following method is a favorite: Pack closely in the barrel, first rubbing the salt well into the exposed ends of bones, and sprinkle well between each layer, using no brine until forty-eight hours after, and then let the brine be strong enough to bear an egg. After six weeks take out the hams and bacon and hang in the smokehouse. When warm weather brings danger of flies, smoke a week with hickory chips; avoid heating the air much. If one has a dark, close smokehouse, the meat can hang in it all summer; otherwise pack in boxes, putting layers of sweet, dry hay between. This method of packing is preferred by some to packing in dry salt or ashes.

RENEWING PORK BRINE.

Not infrequently from insufficient salting and unclean barrels, or other cause, pork placed in brine begins to spoil, the brine smells bad, and the contents, if not soon given proper attention, will be unfit for food. As soon as this trouble is discovered, lose no time in removing the contents from the barrel, washing each piece of meat separately in clean water. Boil the brine for half an hour, frequently removing the scum and impurities that will rise to the surface. Cleanse the barrel thoroughly by washing with hot water and hard wood ashes. Replace the meat after sprinkling it with a little fresh salt, putting the purified brine back when cool, and no further trouble will be experienced, and if the work be well done, the meat will be sweet and firm. Those who pack meat for home use do not always remove the blood with salt. After meat is cut up it is better to lie in salt for a day and drain before being placed in the brine barrel.

A HANDY SALTING BOX.

A trough made as shown at Fig. 16 is very handy for salting meats, such as hams, bacon and beef, for drying. It is made of any wood which will not flavor the meat; ash, spruce or hemlock plank, one and a half inches thick, being better than any others. A good size is four feet long by two and one-half wide and one and one-half deep. The joints should be made tight with white lead spread upon strips of cloth, and screws are vastly better than nails to hold the trough together.

CHAPTER IX.

CARE OF HAMS AND SHOULDERS.

In too many instances farmers do not have the proper facilities for curing hams, and do not see to it that such are at hand, an important point in success in this direction. A general cure which would make a good ham under proper conditions would include as follows: To each 100 lbs. of ham use seven and a half pounds Liverpool fine salt, one and one-half pounds granulated sugar and four ounces saltpeter. Weigh the meat and the ingredients in the above proportions, rub the meat thoroughly with this mixture and pack closely in a tierce. Fill the tierce with water and roll every seven days until cured, which in a temperature of 40 to 50 degrees would require about fifty days for a medium ham. Large hams take about ten days more for curing. When wanted for smoking, wash the hams in water or soak for twelve hours. Hang in the smokehouse and smoke slowly forty-eight hours and you will have a very good ham. While this is not the exact formula followed in big packing houses, any more than are other special recipes given here, it is a general ham cure that will make a first-class ham in every respect if proper attention is given it.

Another method of pickling hams and shoulders, preparatory to smoking, includes the use of molasses. Though somewhat different from the above formula, the careful following of directions cannot fail to succeed admirably. To four quarts of fine salt and two ounces of pulverized saltpeter, add sufficient molasses to make a pasty mixture. The hams having hung in a dry, cool place for three or four days after cutting up, are to be covered all over with the mixture, more thickly on the flesh side, and laid skin side down for three or four days. In the meantime, make a pickle of the following proportions, the quantities here named being for 100 lbs. of hams. Coarse salt, seven pounds; brown sugar, five pounds; saltpeter, two ounces; pearlash or potash, one-half ounce; soft water, four gallons. Heat gradually and as the skim rises remove it. Continue to do this as long as any skim rises, and when it ceases, allow the pickle to cool. When the hams have remained the proper time immersed in this mixture, cover the bottom of a clean, sweet barrel with salt about half an inch deep. Pack in the hams as closely as possible, cover them with the pickle, and place over them a follower with weights to keep them down. Small hams of fifteen pounds and less, also shoulders, should remain in the pickle for five weeks; larger ones will require six to eight weeks, according to size. Let them dry well before smoking.

WESTPHALIAN HAMS.

This particular style has long been a prime favorite in certain markets of Europe, and to a small extent in this country also. Westphalia is a province of Germany in which there is a large industry in breeding swine for the express purpose of making the most tender meat with the least proportion of fat. Another reason for the peculiar and excellent qualities which have made Westphalian hams so famous, is the manner of feeding and growing for the hams, and finally the preserving, curing, and last of all, smoking the hams. The Ravensberg cross breed of swine is a favorite for this purpose. They are rather large animals, having slender bodies, flat groins, straight snouts and large heads, with big, overhanging ears. The skin is white, with straight little bristles.

A principal part of the swine food in Westphalia is potatoes; these are cooked and then mashed in the potato water. The pulp thus obtained is thoroughly mixed with wheat bran in a dry, raw state; little corn is used. In order to avoid overproduction of fat and at the same time further the growth of flesh of young pigs, some raw cut green feed, such as cabbage, is used; young pigs are also fed sour milk freely. In pickling the hams they are first vigorously rubbed with saltpeter and then with salt. The hams are pressed in the pickling vat and entirely covered with cold brine, remaining in salt three to five weeks. After this they are taken out of the pickle and hung in a shady but dry and airy place to "air-dry." Before the pickled hams can be put in smoke they are exposed for several weeks to this drying in the open air. As long as the outside of the ham is not absolutely dry, appearing moist or sticky, it is kept away from smoke.

Smoking is done in special large chambers, the hams being hung from the ceiling. In addition to the use of sawdust and wood shavings in making smoke, branches of juniper are often used, and occasionally beech and alder woods; oak and resinous woods are positively avoided. The smoking is carried on slowly. It is recommended to smoke for a few days cautiously, that is, to have the smoke not too strong. Then expose the hams for a few days in the fresh air, repeating in this way until they are brown enough. The hams are actually in smoke two or three weeks, thus the whole process of smoking requires about six weeks. Hams are preserved after their smoking in a room which is shady, not accessible to the light, but at the same time dry, cool and airy.

THE PIG AND THE ORCHARD.

The two go together well. The pig stirs up the soil about the trees, letting in the sunshine and moisture to the roots and fertilizing them, while devouring many grubs that would otherwise prey upon the fruit. But many orchards cannot be fenced and many owners of fenced orchards, even, would like to have the pig confine his efforts around the trunk of each tree. To secure this have four fence panels made and yard the pig for a short time in succession about each tree, as suggested in the diagram, Fig. 17.

CHAPTER X.

DRY SALTING BACON AND SIDES.

For hogs weighing not over 125 or 130 lbs. each, intended for dry curing, one bushel fine salt, two pounds brown sugar and one pound saltpeter will suffice for each 800 lbs. pork before the meat is cut out; but if the meat is large and thick, or weighs from 150 to 200 lbs. per carcass, from a gallon to a peck more of salt and a little more of both the other articles should be taken. Neither the sugar nor the saltpeter is absolutely necessary for the preservation of the meat, and they are often omitted. But both are preservatives; the sugar improves the flavor of the bacon, and the saltpeter gives it greater firmness and a finer color, if used sparingly. Bacon should not be so sweet as to suggest the "sugar-cure;" and saltpeter, used too freely, hardens the tissues of the meat, and renders it less palatable. The quantity of salt mentioned is enough for the first salting. A little more

NEW SALT IS ADDED AT THE SECOND SALTING

and used together with the old salt that has not been absorbed. If sugar and saltpeter are used, first apply about a teaspoonful of pulverized saltpeter on the flesh side of the hams and shoulders, and then taking a little sugar in the hand, apply it lightly to the flesh surface of all the pieces. A tablespoonful is enough for any one piece.

If the meat at the time of salting is moist and yielding to the touch, rubbing the skin side with the gloved hand, or the "sow's ear," as is sometimes insisted on, is unnecessary; the meat will take salt readily enough without this extra labor. But if the meat is rigid, and the weather very cold, or if the pieces are large and thick, rubbing the skin side to make it yielding and moist causes the salt to penetrate to the center of the meat and bone. On the flesh side it is only necessary to sprinkle the salt over all the surface. Care must be taken to get some salt into every depression and into the hock end of all joints. An experienced meat salter goes over the pieces with great expedition. Taking a handful of the salt, he applies it dextrously by a gliding motion of the hand to all the surface, and does not forget the hock end of the bones where the feet have been cut off. Only dry salt is used in this method of curing. The meat is never put into brine or "pickle," nor is any water added to the salt to render it more moist.

BEST DISTRIBUTION OF THE SALT.

A rude platform or bench of planks is laid down, on which the meat is packed as it is salted. A boy hands the pieces to the packer, who lays down first a course of middlings and then sprinkles a little more salt on all the places that do not appear to have quite enough. Next comes a layer of shoulders and then another layer of middlings, until all these pieces have been laid. From time to time a little more salt is added, as appears to be necessary. The hams are reserved for the top layer, the object being to prevent them from becoming too salt. In a large bulk of meat the brine, as it settles down, lodges upon the lower pieces, and some of them get rather more than their quota of salt. Too much saltiness spoils the hams for first-class bacon. In fact, it spoils any meat to have it too salt, but it requires less to spoil the hams, because, as a rule, they are mostly lean meat. The jowls, heads and livers, on account of the quantity of blood about them, are put in a separate pile, after being salted. The chines and spareribs are but slightly salted and laid on top of the bulk of neat meat. The drippings of brine and blood from the meat are collected in buckets and sent to the compost heaps. If there are rats, they must be trapped or kept out in some way. Cats, also, should be excluded from the house. Close-fitting boxes, which some use to keep the rats from the meat, are not the best; the meat needs air.

In ten days to three weeks, according to weather and size of the meat, break bulk and resalt, using the old salt again, with just a little new salt added. In four to six weeks more, or sooner, if need be, break up and wash the meat nicely, preparatory to smoking it. Some farmers do not wash the salt off, but the meat receives smoke better and looks nicer, if washed.

CURING PORK FOR THE SOUTH.

This requires a little different treatment. It is dry-salted and smoked. The sides, hams and shoulders are laid on a table and rubbed thoroughly with salt and saltpeter (one ounce to five pounds of salt), clear saltpeter being rubbed in around the ends of the bones. The pieces are laid up, with salt between, and allowed to lie. The rubbing is repeated at intervals of a week until the meat is thoroughly salted through, and it is then smoked. It must afterward be left in the smokehouse, canvased or buried in a box of ashes, to protect it from the flies.

CHAPTER XI.

SMOKING AND SMOKEHOUSES.

For best quality of bacon, the proper meat is of first importance. Withes or strings of basket wood, bear's grass, or coarse, stout twine, one in the hock end of each ham and shoulder, and two in the thick side of each middling, are fastened in the meat by which to suspend it for smoking. Before it is hung up the entire flesh surface of the hams and shoulders, and sometimes the middlings also, is sprinkled thickly with fine black pepper, using a large tin pepper box to apply it. Sometimes a mixture of about equal parts of black and red pepper helps very much to impart a good flavor to the meat. It was thought formerly that black pepper, applied to meat before smoking it, would keep the bacon bug (Dermestes) "skippers" from being troublesome. But it is now known that the skipper skips just as lively where the pepper is. The meat is hung upon sticks or on hooks overhead very close together, without actually touching, and is ready for smoking.

THE SMOKEHOUSE.

The meat house is of course one with an earth, brick, or cement floor, where the fire for the smoke is made in a depression in the center of the room, so as to be as far as possible from the walls. A few live coals are laid down, and a small fire is made of some dry stuff. As it gets well to burning, the fire is smothered with green hickory or oak wood, and a basket of green chips from the oak or hickory woodpile is kept on hand and used as required to keep the fire smothered so as to produce a great smoke and but little blaze. If the chips are too dry they are kept wet with water. Care is taken not to allow the fire to get too large and hot, so as to endanger the meat hung nearest to it. Should the fire grow too strong, as it sometimes will, a little water is thrown on, a bucketful of which is kept always on hand. The fire requires constant care and nursing to keep up a good smoke and no blaze. Oak and hickory chips or wood impart the best color to meat. Some woods, as pine, ailanthus, mulberry and persimmon, are very objectionable, imparting a disagreeable flavor to the bacon. Corn cobs make a good smoke for meat, but they must be wet before laying them on the fire. Hardwood sawdust is sometimes advantageously used in making a fire for smoking meats. No blaze is formed, and if it burns too freely can be readily checked by sprinkling a little water upon it. This is a popular method in parts of Europe, and in that country damp wheat straw is also sometimes used to some extent.

COMBINED SMOKEHOUSE AND OVEN.

The oven, shown in Fig. 18, occupies the front and that part of the interior which is represented in our illustration by the dotted lines. The smokehouse occupies the rear, and extends over the oven. The advantages of this kind of building are the perfect dryness secured, which is of great importance in preserving the meat, and the economy in building the two together, as the smoke that escapes from the oven may be turned into the smokehouse. This latter feature, however, will not commend itself to many who prefer the use of certain kinds of fuel in smoking which are not adapted to burning in a bake oven.

Cloudy and damp days are the best for smoking meat. It seems to receive the smoke more freely in such weather, and there is also less danger of fire. The smoke need not be kept up constantly, unless one is in a hurry to sell the meat. Half a day at a time on several days a week, for two or three weeks, will give the bacon that bright gingerbread color which is generally preferred. It should not be made too dark with smoke. It is a good plan, after the meat is smoked nearly enough, to smoke it occasionally for half a day at a time all through the spring until late in May. It is thought that smoke does good in keeping the Dermestes out of the house. The work of smoking may be finished up in a week, if one prefers, by keeping up the smoke all day and at night until bedtime. Some smoke more, others less, according to fancy as to color. No doubt, the more it is smoked, the better the bacon will keep through the summer. But it need not, and, in fact, should not, be made black with smoke.

It is necessary, before the smoking is quite completed, to remove the meat that is in the center just over the fire to one side, and to put the pieces from the sides in the center. The meat directly over the smoke colors faster than that on the sides, although the house is kept full of smoke constantly. Some farmers do not care to risk the safety of their meat by having an open fire under it, and so set up an old stove, either in the room or on the outside, in which latter case a pipe lets the smoke into the house. A smoldering fire is then kept up with corn cobs or chips. But there is almost as much danger this way as the other. The stovepipe may become so hot as to set fire to the walls of the house where it enters, or a blaze may be carried within if there is too much fire in the stove. There is some risk either way, but with a properly built smokehouse, there is no great danger from the plan described.

THE MEAT IS NOW CURED

and, if these directions have been observed, the farmer has a supply of bacon as good as the world can show. Some may prefer a "shorter cut" from the slaughter pen to the baking pan, and with their pyroligenous acid may scout the old-fashioned smoke as heathenish, and get their bacon ready for eating in two hours after the salt has struck in. But they never can show such bacon by their method as we can by ours. There is but one way to have this first-class bacon and ham, and that way is the one herein portrayed.

TO MAKE A SMOKEHOUSE FIREPROOF

as far as the stove ashes are concerned, is not necessarily an expensive job; all that is required is to lay up a row of brick across one end, also two or three feet back upon each side, connecting the sides with a row across the building, making it at least two feet high. As those who have a smokehouse use it nearly every year, that part can also be made safe from fire by the little arch built at the point shown in the illustration, Fig. 19. The whole is laid up in mortar, and to add strength to the structure an iron rod or bar may be placed across the center of the bin and firmly imbedded in the mortar, two or three rows of brick from the top. Of course, the rear of the arch is also bricked up. In most cases, less than 250 brick will be all that is required.

A WELL ARRANGED SMOKEHOUSE.

A simple but satisfactory smokehouse is shown in the illustration, Fig. 20, and can be constructed on the farm at small cost. It is so arranged as to give direct action of smoke upon the meat within, and yet free from the annoyance that comes from entering a smoke-filled room to replenish the fire. The house is square, and of a size dependent upon the material one may have yearly to cure by smoke. For ordinary use, a house ten feet square will be ample. There are an entrance door on one side and a small window near the top that can be opened from the outside to quickly free the inside from the smoke when desired. At the bottom of one side is a small door, from which extends a small track to the center of the room. Upon this slides a square piece of plank, moved by an iron rod with a hook on one end. On the plank is placed an old iron kettle, Fig. 21, with four or five inches of earth in the bottom, and upon this is the fire to be built. The kettle can be slid to the center of the room with an iron rod and can be drawn to the small door at any time to replenish the fire without entering the smoky room or allowing the smoke to come out. The house has an earthen floor and a tight foundation of stone or brick. The walls should be of matched boarding and the roof shingled. The building is made more attractive in appearance if the latter is made slightly "dishing."

SMOKING MEATS IN A SMALL WAY.

A fairly good substitute for a smokehouse, where it is desired to improvise something for temporary use in smoking hams or other meat, may be found in a large cask or barrel, arranged as shown in the engraving, Fig. 22. To make this effective, a small pit should be dug, and a flat stone or a brick placed across it, upon which the edge of the cask will rest. Half of the pit is beneath the barrel and half of it outside. The head and bottom may be removed, or a hole can be cut in the bottom a little larger than the portion of the pit beneath the cask. The head or cover is removed, while the hams are hung upon cross sticks. These rest upon two cross bars, made to pass through holes bored in the sides of the cask, near the top. The head is then laid upon the cask and covered with sacks to confine the smoke. Some coals are put into the pit outside of the cask, and the fire is fed with damp corn cobs, hardwood chips, or fine brush. The pit is covered with a flat stone, by which the fire may be regulated, and it is removed when necessary to add more fuel.

ANOTHER BARREL SMOKEHOUSE.

For those who have only the hams and other meats from one or two hogs to smoke, a practicable smokehouse, like that shown in Fig. 23, will serve the purpose fairly well. A large barrel or good-sized cask should be used, with both heads removed. A hole about a foot deep is dug to receive it, and then a trench of about the same depth and six or eight feet long, leading to the fireplace. In this trench can be laid old stovepipe and the ground filled in around it. The meat to be smoked is suspended in the barrel and the lid put on, but putting pieces under it, so there will be enough draft to draw the smoke through. By having the fire some distance from the meat, one gets the desired amount of smoke and avoids having the meat overheated.

CHAPTER XII.

KEEPING BACONS AND HAMS.

The ideal meat house or smokehouse is a tall frame structure, twelve by fifteen or fifteen by eighteen feet, underpinned solidly with brick set a foot or more into the ground, or with a double set of sills, the bottom set being buried in the soil. This mode of underpinning is designed to prevent thieves from digging under the wall and into the house. Stout, inch-thick boards are used for the weatherboarding, and sometimes the studs are placed near enough together to prevent a person from getting through between them. The house is built tall to give more room for meat and to have it farther from the fire while it is being smoked. The weatherboarding and the roof should be tight to prevent too free escape of the smoke. No window, and but one door, is necessary. The floor should be of clay, packed firm, or else laid in cement or brick. Indeed, it would be better to have the entire walls built of brick, but this would add considerably to the cost of construction.

THE ROOM SHOULD BE LARGE ENOUGH

to admit of a platform on one or both sides, upon which to pack the pork when salted. There should be a salt barrel, a large wooden tray made of plank, in which to salt the meat, and a short, handy ladder for reaching the upper tier of joists. A large basket for holding chips, a tub for water when smoking meat, a large chopping block and a meat axe, for the convenience of the cook, are necessary articles for the meat house. Nothing else should be allowed to cumber the room to afford a harbor for rats or to present additional material for a blaze, in case a spark from the fire should snap out to a distance. The house should be kept neatly swept, and rats should not be allowed to make burrows under anything in the room. The floor of the meat house should always be of some hard material like cement or brick, or else clay pummeled very hard, so that there would be no hiding place for the pupae of the Dermestes (parent of the "skipper").

The skipper undergoes one or two moltings while in the meat, and at last drops from the bacon to the floor, where, if the earth is loose, it burrows into the ground and, remaining all winter, comes out a perfect beetle in spring. A hard, impervious floor will prevent it from doing this, and compel it to seek a nesting place elsewhere. The reason why country bacon is sometimes so badly infested with the skipper is that the house and floor afford or become an excellent incubator, as it were, for the Dermestes, and the bacon bugs become so numerous that all the meat gets infested with them. In case the floor of the smokehouse is soft and yielding, it becomes necessary each winter, before the meat is packed to salt, to remove about two inches of the soil and put in fresh earth or clay in its place. Thus, many of the insects would be carried out, where they would be destroyed. The walls and roof of the room on the interior should also be swept annually to dislodge any pupae that might be hibernating in the cracks and crevices. With these precautions, there should not be many of the pests left within the building, though it is a hard matter when a house once gets badly infested to dislodge them entirely. There are so many hiding places about a plain shingle roof that it is next to impossible not to have some of these insects permanently lodged in the meat house. But with a good, hard floor, frequent sweeping and the use of plenty of black pepper on the meat, the number of the Dermestes should be reduced to the minimum.

BACON KEEPS NOWHERE SO WELL

as in the house where it is smoked, and if the bugs do not get too numerous it is decidedly better to allow it to remain hanging there. Bacon needs air and a cool, dry, dark room for keeping well in summer. The least degree of dampness is detrimental, causing the bacon to mold. It has been noticed, however, that moldy bacon is seldom infested with the skipper. Hence some people, to keep away the skippers, hang their bacon in a cellar where there is dampness, preferring to have it moldy rather than "skippery." Some housekeepers preserve hams in close boxes or barrels, in a cool, dark room, and succeed well. Others pack in shelled oats or bran, or wrap in old newspapers and lay away on shelves or in boxes. Inclosing in cloth sacks and painting the cloth is also practiced. All these plans are more or less successful, but oblige the housekeeper to be constantly on the watch to prevent mice and ants from getting to the bacon. But if anyone should prefer

TO EXCLUDE THE BUGS ENTIRELY

from his meat the following contrivance is offered as a cheap and entirely satisfactory arrangement: After the meat is thoroughly smoked, hang all of it close together, or at least all the hams, in the center of the house, and inclose it on all sides with a light frame over which is stretched thin cotton cloth, taking care that there shall be no openings in the cloth or frame through which the bugs might crawl. There let it hang all summer. This contrivance will prevent the bug from getting at the meat to deposit its eggs, and the thin, open fabric of the cloth will at the same time admit plenty of air. The bottom or one side of the frame should be fixed upon hinges, for convenience in getting at the bacon as wanted. As the bacon bug comes out in March, or April farther south, in February it is necessary to get the meat smoked and inclosed under the canvas before the bug leaves its winter quarters. Hams may be thus kept in perfect condition as long as may be desirable, and will remain sweet and nice many months.

BOX FOR STORING BACON.

If the smokehouse is very dark and close, so that the flies or bugs will not be tempted to or can get in, all that is necessary is to have the meat hung on the pegs; but, if not, even when the meat is bagged, there is still some risk of worms. To provide a box that will be bugproof, ratproof, and at the same time cool, as seen in the illustration, Fig. 24, make a frame one inch thick and two or three inches wide, with a close plank bottom; cover the whole box with wire cloth, such, as is used for screens. Let the wire cloth be on the outside, so that the meat will not touch it. The top may be of plank and fit perfectly tight, so that no insect can creep under. Of course, the box may be of any size desired. It will be well to have the strips nailed quite closely together, say, about one and a half inches apart. When the meat is put in, lay sticks between, so that the pieces will not touch. If the box is made carefully, it is bugproof and ratproof, affording ventilation at the same time, and so preventing molding. Meat should be kept in a dry and cool place.

CHAPTER XIII.

SIDELIGHTS ON PORK MAKING.

The trade in country dressed hogs varies materially from year to year. Since the big packing houses have become so prominent in the industry there is, of course, less done in country dressed hogs, yet a market is always found for considerable numbers. Thirty years ago Chicago received as many as 350,000 dressed hogs in one year. With a growth of the packing industry this business decreased, until 1892, when only 5000 were handled at Chicago, but since that date there has been a revival of interest, with as many as 60,000 received in 1894 and an ever changing number since that date. Thirty years ago the number of hogs annually packed at Chicago was about 700,000. This business has increased since to as many as 8,000,000 in a year, the industry in other packing centers being in much the same proportion. At all packing centers in the west there are slaughtered annually 20,000,000 to 24,000,000 hogs.

Compared with the enormous numbers fattened and marketed on the hoof, a very small proportion of the hogs turned off the farms each year are sold dressed. Yet with many farmers, particularly those who have only a small number to dispose of, it is always a question as to which is the better way to sell hogs, dressed or alive. No individual experience can be taken as a criterion, yet here is a record of what one Michigan farmer did in the way of experiment. He had two lots of hogs to sell. One litter of seven weighed a total of 1605 lbs. alive, and dressed 1,335 lbs., which was three pounds over a one-sixth shrinkage; one litter of five weighed 1540 lbs. and dressed 1320 lbs., losing exactly one-seventh, they being very fat. The sow weighed 517 lbs. and dressed 425, dressing away about 18 lbs. to the 100 lbs. He was offered $3.80 per 100 lbs. live weight, for all the hogs, and $3 for the sow. He finally sold the seven hogs, dressed, at $5 per 100 lbs., the second lot of five at $4.75, and the sow at $4.25. He decided that by dressing the hogs before selling, he gained about $12.50, aside from lard and trimmings. The experience here noted would not necessarily hold good anywhere and any time. Methods employed in packing hogs have been brought down to such a fine point, however, with practically every portion utilized, that unless a farmer has a well-defined idea where he can advantageously sell his dressed pork, it would not pay, as a general thing, to butcher any considerable number of hogs, with a view of thus disposing of them.

AN EASILY FILLED PIG TROUGH.

To get swill into a pig trough is no easy matter if the hogs cannot be kept out until it is filled. The arrangement shown in Fig. 25 will be found of much value and a great convenience. Before pouring in the swill, the front end of the pen, in the form of a swinging door suspended from the top, is placed in the position shown at b. The trough is filled and the door allowed to assume the position shown at a.

AN AID IN RINGING HOGS.

A convenient trap for holding a hog while a ring is placed in its nose consists of a trunk or a box without ends, 6 feet long, 30 inches high and 18 inches wide, inside measure. This trunk has a strong frame at one end, to which the boards are nailed. The upper and lower slats are double, and between them a strong lever has free play. To accommodate large or small pigs, two pins are set in the lower slat, against which the lever can bear. The pins do not go through the lever. This trunk is placed in the door of the pen, and two men are required to hold it and ring the hogs. When a hog enters and tries to go through, one man shoves the lever up, catching him just back of the head, and holds him there. The second man then rings him, and he is freed. Fig. 26 exhibits the construction of the trap, in the use of which one can hold the largest hog with ease.

AVERAGE WEIGHTS OF LIVE HOGS.

The average weight of all hogs received at Chicago in 1898 was 234 lbs.; in 1896, 246 lbs. The average weight of all hogs received at Chicago in 1895 was 230 lbs.; in 1894, 233 lbs.; in 1893, 240 lbs.

EXTREMES IN MARKET PRICE OF PORK AND LARD.

The highest price of mess pork at Chicago during the last forty years, according to the Daily Trade Bulletin, was $44 per bbl. in 1864, and the lowest price $5.50 per bbl., paid in 1896. The highest price of lard was naturally also in war times, 30c per lb. in 1865; the lowest price a shade more than 3c, in 1896.

NET TO GROSS.

Good to prime hogs, when cut up into pork, hams, shoulders and lard, will dress out 73 to 75 per cent, according to the testimony of the large packing concerns. That is, for every 100 lbs. live weight, it is fair to estimate 73 to 75 lbs. of product of the classes named. If cut into ribs instead of pork, prime hogs would net 70 to 72 per cent, while those which are not prime run as low as 65 per cent. For comparative purposes, it may be well to note here that good farm-fed cattle will dress 54 to 56 per cent of their live weight in beef, the remainder being hide, fat, offal, etc., and sheep will dress 48 to 54 per cent, 50 per cent being a fair average.

RELATIVE WEIGHTS OF PORTIONS OF CARCASS.

To determine the relation of the different parts of the hog as usually cut, to the whole dressed weight, the Alabama experiment station reports the following results. The test was made with a number of light hogs having an average dressed weight of 137 lbs. The average weight of head was 12.2 lbs.; backbone, 6.8 lbs.; the two hams, 25.4 lbs.; the two shoulders, 33.1 lbs.; leaf lard, 8.3 lbs.; ribs, 5.5 lbs.; the two "middling" sides, 35 lbs.; tender loin, 1.1 lbs.; feet, 3.6 lbs.

GATES FOR HANDLING HOGS.

The device shown in the accompanying illustrations for handling hogs when they are to be rung or for other purposes, is very useful on the ordinary farm. Fig. 27 represents a chute and gate which will shut behind and before the hog and hold him in position. There is just room enough for him to stick his nose out, and while in this position rings can be inserted. The sides of the chute must be much closer together than shown in the engraving, so that the hog cannot turn about. In fact, the width should be just sufficient to allow the hog to pass through. Fig. 28 represents the side view of another gate and pen, so arranged that the door can be opened and shut without getting into the pen.

CHAPTER XIV.

PACKING HOUSE CUTS OF PORK.

While considering primarily the proper curing of pork for use on the farm and for home manufacture by farmers, it will not be out of the way to become acquainted with some of the leading cuts of meat as made by the big pork packers at Chicago and elsewhere. In the speculative markets, a large business is done in "mess pork," "short ribs" and lard. These are known as the speculative commodities in pork product. The prices established, controlled largely by the amount offered and the character of the demand, regulate to a considerable extent the market for other cuts of pork, such as long clear middles, hams and shoulders. Our illustrations of some of the leading cuts of meats, furnished us through the courtesy of Hately Bros., prominent pork packers in Chicago, together with accompanying descriptions, give a very good idea of the shape pork product takes as handled in the big markets of the world.

MESS PORK.

This standard cut, Fig. 29, is made from heavy fat hogs. The hog is first split down the back, the backbone being left on one side. Ham and shoulders taken off, the sides are then cut in uniform strips of four or five pieces. Equal portions of both sides are then packed in barrels, 200 lbs. net, the pieces numbering not more than sixteen nor less than nine. Barrels to be filled with a pickle made with 40 lbs. of salt to each barrel.

SHORT RIBS.

These are made from the sides, with the ham and shoulder taken off and backbone removed; haunchbone and breastbone sawed or cut down smooth and level with the face of the side. The pieces (Fig. 30) are made to average 32 lbs. and over.

SHOULDERS.

Regular shoulders (Fig. 31), or commonly called dry salted shoulders, are cut off the sides between first and second ribs, so as not to expose forearm joint. Shank cut off at knee joint. Neck bone taken out and neck trimmed smooth. Shoulders butted off square at top. Made to average 12 to 14, 14 to 16, and 16 to 18 lbs. On the wholesale markets can usually be bought at about the price per pound of live hogs.

HAMS.

American cut hams are cut short inside the haunchbone, are well rounded at butt and all fat trimmed off the face of the hams to make as lean as possible. See Fig. 32. Cut off above the hock joint. Hams are made to average 10 to 12, 12 to 14, 14 to 16, 16 to 18, and 18 to 20 lbs.

PICNIC HAMS.

This is a contradictory term, for the picnic ham is in truth a shoulder. Picnic hams (Fig. 33) are made from shoulders cut off sides between second and third ribs. Shank bone cut off one inch above knee joint, and neck bone taken out. Butt taken off through the middle of the blade and nicely rounded to imitate a ham. Made to average 5 to 6, 6 to 8, 8 to 10, and 10 to 14 lbs.

WILTSHIRE CUT BACON.

This cut (Fig. 34) is from hogs weighing about 150 lbs. Formerly the hair was removed by singeing, but this method is not so much employed now. The Wiltshire bacon is consumed almost entirely in London, Bristol and the south of England generally.

STANDARD LARD.

The following is the rule in force at Chicago for the manufacture of standard prime steam lard: Standard prime steam lard shall be solely the product of the trimmings and other fat parts of hogs, rendered in tanks by the direct application of steam and without subsequent change in grain or character by the use of agitators or other machinery, except as such change may unavoidably come from transportation. It must have proper color, flavor and soundness for keeping, and no material which has been salted must be included. The name and location of the renderer and the grade of the lard shall be plainly branded on each package at the time of packing.

NEUTRAL LARD.

This is made at the big packing houses from pure leaf lard, which after being thoroughly chilled is rendered in open tanks at a temperature of about 120 degrees. The portion rendered at this temperature is run into packages and allowed to cool before closing tightly.

Lard stearine is made from the fat of hogs which is rendered and then pressed and the oil extracted. The oil is used for lubricating purposes, and the stearine by lard refiners in order to harden the lard, especially in warm weather.

CHAPTER XV.

MAGNITUDE OF THE SWINE INDUSTRY.

Were it not for the foreign demand for our pork and pork product there would be much less profit in fattening hogs for market than there is, irrespective of the price of corn and other feeds. England is our best customer, taking by far the larger part of our entire exports of all lard, cured meats and other hog product, but there is an encouraging trade with other foreign countries. The authorities at Washington are making every effort to enlarge this foreign outlet. Certain European countries, notably France and Germany, place irksome embargoes on American pork product. Ostensibly, these foreign governments claim the quality and healthfulness of some of the American pork are in question, but in reality back of all this is the demand from the German and French farmers that the competition afforded by American pork must be kept down. It is believed that eventually all such restrictions will be swept away, through international agreement, and that thus our markets may be further extended, greatly benefiting the American farmer. Our exports of hog product, including pork, bacon, hams and lard, represent a value annually of about $100,000,000.

THE WORLD'S SUPPLY OF BACON

is derived chiefly from the United States, which enjoys an enormous trade with foreign consuming countries, notably England and continental Europe. Irish bacon is received with much favor in the English markets, while Wiltshire and other parts of England also furnish large quantities, specially cured, which are great favorites among consumers. Some idea of the magnitude of the foreign trade of the United States, so far as hog product is concerned, may be formed by a glance at the official figures showing our exports in a single year. During the twelve months ended June 30, 1899, the United States exported 563,000,000 lbs. bacon, 226,000,000 lbs. ham, 137,000,000 lbs. pickled pork and 711,000,000 lbs. lard, a total of 1,637 million pounds pork product. On the supposition that live hogs dress out, roughly speaking, 70 per cent product, this suggests the enormous quantity of 2,340 million pounds of live hogs taken for the foreign trade in one year. Estimating the average weight at 240 lbs., this means nearly 10,000,000 hogs sent to American slaughterhouses in the course of one year to supply our foreign trade with pork product. The United Kingdom is by far our best customer, although we export liberal quantities to Belgium, Holland, Germany, France, Canada, Brazil, Central America and the West Indies. Total value of our 1899 exports of pork product was $116,000,000.

The enormous business of the big packing houses, located chiefly in the west, with a few in the east, can scarcely be comprehended in its extent. Chicago continues to hold the prestige of the largest packing center in the world, but other western cities are crowding it. In 1891 Chicago received 8,600,000 hogs, the largest on record, most of which were packed in that city, and the product shipped all over the world. In recent years the Chicago receipts have averaged smaller, but the proportion going to the packing concerns remains about the same. It is estimated that the hogs received at that city in 1898 had a value of $84,000,000.

CO-OPERATIVE CURING HOUSES IN DENMARK.

About half the pork exported to England from Denmark is cured by the co-operative curing houses, established first in 1888 and since that date greatly increased in number. Enormous quantities of cheap Black Sea barley have been brought into Denmark the last few years, used principally for fodder. The principal advantage of the co-operative system, doing away with the middleman, applies to these establishments. Farmers who raise hogs in a given district of say ten to twenty miles' circumference, unite and furnish the money necessary for the construction and operation of the co-operative curing establishment. The farmers bind themselves to deliver all hogs that they raise to the curing house, and severe fines are collected when animals are sold elsewhere. At every curing house there is a shop for the sale of sausage, fat, etc., these as a rule paying well and forming an important part of the profits in this co-operation.

HOG PRICES AT CHICAGO, PER 100 POUNDS.

Heavy packing, Mixed packing, Light bacon. Year. 260 to 450 lbs. 200 to 250 lbs. 150 to 200 lbs.

1899 $3.10@4.75 $3.50@5.00 $3.75@5.00 1898 3.25@4.80 3.30@4.75 3.00@4.65 1897 3.00@4.50 3.20@4.60 3.20@4.65 1896 2.40@4.45 2.75@4.45 2.80@4.45 1895 3.25@5.45 3.25@5.55 3.25@5.70 1894 3.90@6.75 3.90@6.65 3.50@6.45 1893 3.80@8.75 4.25@8.65 4.40@8.50 1892 3.70@7.00 3.65@6.70 3.60@6.85 1891 3.25@5.70 3.25@5.75 3.15@5.95

TOTAL PACKING AND MARKETING OF HOGS.

[Year Ended March 1--Cincinnati Price Current.]

Receipts. Western Eastern N. Y., Phil. Packing. Packing. and Balto. Total.

1898-99 23,651,000 3,162,000 2,978,000 29,791,000 1897-98 20,201,000 3,072,000 2,861,000 26,134,000 1896-97 16,929,000 2,791,000 2,950,000 22,670,000 1895-96 15,010,000 2,603,000 2,867,000 20,480,000 1894-95 16,003,000 3,099,000 2,517,000 21,619,000 1893-94 11,605,000 2,701,000 2,483,000 16,789,000 1892-93 12,390,000 3,016,000 2,790,000 18,196,000 1892 14,457,000 2,771,000 3,684,000 20,912,000 1891 17,713,000 2,540,000 3,713,000 23,966,000

RECEIPTS OF HOGS AT LEADING POINTS BY YEARS.

[Stated in thousands--From American Agriculturist Year Book for 1898.]

1897 1896 1895 1894 1893 1892 1891

Chicago 8,364 7,659 7,885 7,483 6,057 7,714 8,601 Kansas City 3,351 2,606 2,458 2,547 1,948 2,397 2,599 Omaha 1,605 1,198 1,188 1,904 1,435 1,706 1,462 St. Louis 1,627 1,618 1,085 1,147 777 848 841 ---- ---- ---- ---- ---- ---- ---- Total 14,947 13,081 12,616 13,081 10,217 12,665 13,503

[1]Cincinnati 875 823 773 639 592 587 816 Indianapolis 1,253 1,255 1,109 964 879 1,123 1,320 Cleveland 750 500 375 295 270 .. .. Detroit 320 224 189 164 149 134 87 ---- ---- ---- ---- ---- ---- ---- Total 3,198 2,802 2,346 1,062 1,890 1,844 2,223

New York 1,578 1,845 1,763 1,656 1,488 1,826 2,177 Boston 1,420 1,435 1,400 1,673 1,373 1,636 1,473 Buffalo 5,621 4,941 5,256 5,626 6,058 6,112 7,167 Pittsburg 1,894 1,340 1,063 1,074 999 1,347 1,428 Philadelphia 250 278 259 280 275 337 377 ---- ---- ---- ---- ---- ---- ---- Total 10,763 9,839 9,741 10,317 10,193 11,258 12,622

St. Paul 225 314 364 327 194 239 263 Sioux City 350 279 341 499 329 413 397 Cedar Rapids 487 358 365 317 293 409 502 St. Joseph, Mo 400 193 252 398 240 289 178 Ft. Worth, Tex 114 141 60 .. .. New Orleans 18 28 26 26 30 36 33 Denver 75 57 48 94 62 83 80 ---- ---- ---- ---- ---- ---- ---- Total 1,669 1,370 1,456 1,661 1,148 1,769 1,453

Montreal 93 89 74 87 70 52 43 Toronto 77 194 154 140 75 74 51 ---- ---- ---- ---- ---- ---- ---- Total 170 283 228 227 145 126 94

1889 1888 1887

Chicago 5,999 4,922 5,471 Kansas City 2,074 2,009 2,423 Omaha 1,207 1,284 1,012 St. Louis 773 652 772 ---- ---- ---- Total 10,053 8,867 9,678

[1]Cincinnati .. .. .. Indianapolis 1,158 896 1,149 Cleveland .. .. .. Detroit 114 21 49 ---- ---- ---- Total .. .. ..

New York 1,762 1,550 1,792 Boston 1,152 1,046 1,047 Buffalo 5,776 5,333 5,074 Pittsburg 1,205 1,161 1,259 Philadelphia 332 281 274 ---- ---- ---- Total 10,247 9,371 9,446

St. Paul 249 273 .. Sioux City 593 431 .. Cedar Rapids 346 307 847 St. Joseph, Mo 253 258 .. Ft. Worth, Tex .. .. .. New Orleans .. .. .. Denver 75 64 54 ---- ---- ---- Total 1,516 1,333 ..

Montreal 23 26 .. Toronto 57 36 35 ---- ---- ---- Total 80 62 35

[1] For year ended March 31.

CRATE FOR MOVING SWINE OR OTHER ANIMALS.

It is often desirable to move a small animal from one building to another, or from one pasture enclosure to another. The illustration, Fig. 35, shows a crate on wheels, with handles permitting it to be used as a wheelbarrow. Into this the pig can be driven, the door closed and the crate wheeled away. It will also be found a very useful contrivance in bringing in calves that have been dropped by their dams in the pasture.

CHAPTER XVI.

DISCOVERING THE MERITS OF ROAST PIG.

By Charles Lamb.

The art of roasting, or rather broiling (which I take to be the elder brother) was accidentally discovered in the manner following. The swineherd, Ho-ti, having gone out into the woods one morning, as his manner was, to collect mast for his hogs, left his cottage in the care of his eldest son, Bo-bo, a great, lubberly boy, who, being fond of playing with fire, as younkers of his age commonly are, let some sparks escape into a bundle of straw, which, kindling quickly, spread the conflagration over every part of their poor mansion, till it was reduced to ashes. Together with the cottage (a sorry, antediluvian makeshift of a building, you may think it), what was of much more importance, a fine litter of new-farrowed pigs, no less than nine in number, perished. China pigs have been esteemed a luxury all over the east, from the remotest periods that we read of. Bo-bo was in the utmost consternation, as you may think, not so much for the sake of the tenement, which his father and he could easily build up again with a few dry branches, and the labor of an hour or two, at any time, as for the loss of the pigs.

While he was thinking what he should say to his father, and wringing his hands over the smoking remnants of one of those untimely sufferers, an odor assailed his nostrils, unlike any scent which he had before experienced. What could it proceed from?--not from the burnt cottage--he had smelt that smell before--indeed, this was by no means the first accident of the kind which had occurred through the negligence of this unlucky firebrand. Much less did it resemble that of any known herb, weed or flower. A premonitory moistening at the same time overflowed his nether lip. He knew not what to think. He next stooped down to feel the pig, if there were any signs of life in it. He burnt his fingers, and to cool them he applied them in his booby fashion to his mouth. Some of the crumbs of the scorched skin had come away with his fingers, and for the first time in his life (in the world's life, indeed, for before him no man had known it), he tasted--crackling!

Again he felt and fumbled at the pig. It did not burn him so much now, still he licked his fingers from a sort of habit. The truth at length broke into his slow understanding, that it was the pig that smelt so, and the pig that tasted so delicious, and, surrendering himself up to the new-born pleasure, he fell to tearing up whole handfuls of the scorched skin with the flesh next it, and was cramming it down his throat in his beastly fashion, when his sire entered amid the smoking rafters, armed with retributory cudgel, and, finding how affairs stood, began to rain blows upon the young rogue's shoulders, as thick as hailstones, which Bo-bo headed not any more than if they had been flies. The tickling pleasure, which he experienced in his lower regions, had rendered him quite callous to any inconveniences that he might feel in those remote quarters. His father might lay on, but he could not beat him from his pig till he had made an end of it, when, becoming a little more sensible of his situation, something like the following dialogue ensued:

"You graceless whelp, what have you got there devouring? Is it not enough that you have burnt me down three houses with your dog's tricks, and be hanged to you! but you must be eating fire, and I know not what--what have you got there, I say?"

"O, father, the pig, the pig! do come and taste how nice the burnt pig eats."

The ears of Ho-ti tingled with horror. He cursed his son, and he cursed himself that ever he should beget a son that should eat burnt pig.

Bo-bo, whose scent was wonderfully sharpened since morning, soon raked out another pig, and fairly rending it asunder, thrust the lesser half by main force into the fists of Ho-ti, still shouting out, "Eat, eat, eat the burnt pig, father, only taste--O Lord!" with suchlike barbarous ejaculations, cramming all the while as if he would choke.

Ho-ti trembled in every joint while he grasped the abominable thing, wavering whether he should not put his son to death for an unnatural young monster, when the crackling scorched his fingers, as it had done his son's, and applying the same remedy to them, he in his turn tasted some of its flavor, which, make what sour mouths he would for a pretence, proved not altogether displeasing to him. In conclusion, both father and son fairly sat down to the mess, and never left off till they had dispatched all that remained of the litter.

CHAPTER XVII.

COOKING AND SERVING PORK.

FIRST PRIZE WINNERS IN THE AMERICAN AGRICULTURIST CONTEST FOR BEST RECIPES FOR COOKING AND SERVING PORK.

PORK PIE.

Unless you have a brick oven do not attempt this dish, as it requires a long and even baking, which no stove oven can give. Make a good pie crust and line a large pan, one holding about 6 quarts; in the bottom put a layer of thin slices of onions, then a layer of lean salt pork, which has been previously browned in the frying pan, next place a layer of peeled apples, which sprinkle with a little brown sugar, using 1/2 lb. sugar to 3 lbs. apples; then begin with onions, which sprinkle with pepper, pork and apples again, and so on until the dish is full. Wet the edges of the crust, put on the top crust, well perforated, and bake at least four hours, longer if possible. These pies are eaten hot or cold and are a great favorite with the English people. Potatoes may be used in place of apples, but they do not give the meat so fine a flavor.

PORK POTPIE.

Three pounds pork (if salt pork is used, freshen it well), cut into inch cubes. Fry brown, add a large onion sliced, and a teaspoon each of chopped sage, thyme and parsley. Cover with 5 pints of water and boil for two hours, add a large pepper cut small or a pinch of cayenne, and a tablespoon of salt if fresh pork has been used. Add also 3 pints vegetables, carrots, turnips and parsnips cut small, boil half an hour longer, when add a pint of potatoes cut into small pieces, and some dumplings. Cover closely, boil twenty minutes, when pour out into a large platter and serve. The dumplings are made of 1 pint of flour, 1 teaspoon salt, and 1 teaspoon baking powder, sifted together. Add 2 eggs, well beaten and 1 cup of milk. Mix out all the lumps and drop by spoonfuls into the stew. Serve this potpie with a salad of dandelion leaves, dressed with olive oil, vinegar, salt and pepper.

PORK GUMBO.

Cut into small dice 2 lbs. lean pork. (In these recipes where the pork is stewed or baked in tomatoes or water, salt pork may be used, provided it is well freshened.) Fry the pork a pale brown, add 2 sliced onions, and when these are brown add 3 bell peppers sliced, and 2 quarts peeled tomatoes, with 2 teaspoons salt. Let boil gently, stirring frequently, for 1-1/2 hours. Peel and cut small 1 pint of young tender okra pods, and add. Cover again and boil half an hour longer. Cook in a lined saucepan, as tin will discolor the okra. With this serve a large dish of rice or hominy. Corn may be used in place of okra if the latter is disliked. The corn should be cut from the cobs and added half an hour before dinner time.

SUCCOTASH.

Boil a piece of lean pork (about 5 lbs. in weight) in 3 quarts water, until the meat is tender. The next day take out the pork, and remove the grease risen on the liquor from the pork during cooking. To 3 pints of the liquor add 1 pint of milk and 1-1/2 pints lima beans. Let them boil until tender--about one hour--when add 1-1/2 pints corn cut from the cob. Let the whole cook for ten minutes, add a teaspoon of salt if necessary, half a teaspoon of pepper, and drop in the pork to heat. When hot, pour into a tureen and serve.

PORK PILLAU.

Take a piece of pork (about 4 lbs.) and 2 lbs. bacon. Wash and put to boil in plenty of water, to which add a pepper pod, a few leaves of sage and a few stalks of celery. One hour before dinner, dip out and strain 2 quarts of the liquor in which the pork is boiling, add to it a pint of tomatoes peeled, a small onion cut fine, and salt if necessary; boil half an hour, when add 1 pint of rice well washed. When it comes to a boil draw to the back of stove and steam until the rice is cooked and the liquor absorbed. The pork must boil three or four hours. Have it ready to serve with the rice. This makes a good dinner, with a little green salad, bread and butter and a good apple pudding.

PORK ROLL.

Chop fine (a meat chopper will do the work well and quickly) 3 lbs. raw lean pork and 1/4 lb. fat salt pork. Soak a pint of white bread crumbs in cold water. When soft squeeze very dry, add to the chopped meat with a large onion chopped fine, 1 tablespoon chopped parsley, 1/2 teaspoon each of chopped sage and thyme, and 1/2 teaspoon black pepper. Mix together thoroughly and form into a roll, pressing it closely and compactly together. Have ready about a tablespoon of fat in a frying pan, dredge the roll thickly with flour and brown it in the fat, turning it until nicely browned on all sides. Then place it in a baking pan, and bake in a hot oven for one hour. Baste it every ten minutes with water. Do not turn or disturb the meat after it has been put into the oven. Half an hour before dinner add 12 or 14 small carrots that have been parboiled in salted boiling water for fifteen minutes. When done, place the roll on a platter, surround it with plain boiled macaroni, dot with the carrots and pour over all a nicely seasoned tomato sauce.

PEPPER POT.

Cut 3 lbs. rather lean pork into 2-inch cubes, fry until brown, place in a 3-quart stone pot (a bean jar is excellent for this purpose) having a close-fitting lid; add 2 large onions sliced, 6 large green peppers (the bell peppers are the best, being fine in flavor and mild), a tablespoon of salt (if fresh pork was used), and 3 large tomatoes peeled and cut small. Fill the pot with water and place in the oven or on the back of the stove and allow to simmer five or six hours, or even longer. The longer it is cooked the better it will be. Persons who ordinarily cannot eat pork will find this dish will do them no harm. The sauce will be rich and nicely flavored, and the meat tender and toothsome. Serve with it plenty of boiled rice or potatoes.

PORK CROQUETTES (IN CABBAGE LEAVES).

To 1 lb. lean pork chopped fine add 1 teaspoon salt, 1/2 teaspoon each of pepper, chopped sage and thyme, 1 teaspoon chopped parsley and a large onion also chopped. Mix well and stir in 2-3 cup (half-pint cup) of well-washed raw rice. Wash a large cabbage, having removed all the defective outer leaves. Plunge it whole into a large pot of boiling salted water and boil for five minutes, remove and drain. This will render the leaves pliable. Let cool a little, when pull the leaves apart, and wrap in each leaf a tablespoon of the pork and rice. Wrap it up securely and neatly as if tying up a parcel and secure with wooden toothpicks or twine. When all are done, lay in a baking dish and cover with a quart of tomatoes peeled and cut fine, mixed with half a pint of water, and a teaspoon of salt. Bake one hour in a hot oven, turning the croquettes occasionally. If the sauce becomes too thick, dilute with a little hot water. When done, dish, pour over the sauce and serve with potatoes or hominy. These are very good indeed. If desired the croquettes may be steamed over hot water in a steamer for three hours, or plunged directly into a kettle of boiling water and boiled for one hour. They are not so delicate as when baked.

PORK WITH PEA PUDDING (ENGLISH STYLE).

Boil the pork as directed above, and do not omit the vegetables, as they flavor the meat and the pudding. Use the yellow split peas and soak a pint in cold water over night. Drain and tie them loosely in a pudding bag and boil with the pork for three hours. An hour before dinner remove and press through a colander, add a teaspoon salt, half a teaspoon pepper and 3 eggs well beaten. Chop enough parsley to make a teaspoonful, add to the peas with a little grated nutmeg. Beat up well, sift in half a pint of flour and pour into a pudding bag. The same bag used before will do if well washed. Tie it up tightly, drop into the pork water again and boil another hour. Remove, let drain in the colander a few minutes, when turn out onto a dish. Serve with the pork, and any preferred sauce; mint sauce is good to serve with pork, and a tomato sauce is always good. In fact, it is a natural hygienic instinct which ordains a tart fruit or vegetable to be eaten with pork. The Germans, who are noted for their freedom from skin diseases, add sour fruit sauces to inordinately fat meats.

PORK WITH SAUERKRAUT (GERMAN STYLE).

Boil a leg of pork for three or four hours, wash 2 quarts sauerkraut, put half of it into an iron pot, lay on it the pork drained from the water in which it was cooking and cover with the remainder of sauerkraut, add 1 quart water in which the pork was cooking, cover closely and simmer gently for one hour.

PORK CHOWDER.

Have ready a quart of potatoes sliced, 2 large onions sliced, and 1 lb. lean salt pork. Cut the pork into thin slices and fry until cooked, drain off all but 1 tablespoon fat and fry the onions a pale brown. Then put the ingredients in layers in a saucepan, first the pork, then onions, potatoes and so on until used, adding to each layer a little pepper. Add a pint of water, cover closely and simmer fifteen minutes, then add a pint of rich milk, and cover the top with half a pound of small round crackers. Cover again and when the crackers are soft, serve in soup plates. If you live where clams are plentiful, add a quart of cleaved clams when the potatoes are almost done and cook ten minutes.

SEA PIE.

Make a crust of 1 quart flour, 2 teaspoons baking powder, 1 teaspoon salt, mix well, rub in a tablespoon of fat--pork fat melted or lard--and mix into a smooth paste with a pint of water. Line a deep pudding dish with this, put in a layer of onions, then potatoes sliced, then a thin layer of pork in slices, more onions, etc., until the dish is full. Wet the edges, put on a top crust. Tie a floured cloth over the top and drop into a pot of boiling water. Let the water come up two-thirds on the dish, and keep the water boiling for four hours. Invert on a dish, remove the mold and serve hot.

For Fresh Pork Only.

CORN AND PORK SCALLOP.

Cut about 2 lbs. young pork into neat chops and reject all fat and bone. Fry them until well cooked and of a pale brown, dust with salt and pepper. Cut some green corn from the cob. Take a 2-quart dish, put a layer of corn in the bottom, then a layer of pork, and so on until the dish is full, add 1 pint of water, cover and bake for one hour. Remove the cover fifteen minutes before serving, so the top may be nicely browned. Serve with potatoes and a lettuce salad. Onions and pork may be cooked in the same manner.

STUFFED SHOULDER OF PORK.

Take a shoulder of pork and bone it. Cut out the shoulder blade, and then the leg bone. After the cut made to extract the shoulder blade, the flesh has to be turned over the bone as it is cut, like a glove-finger on the hand; if any accidental cut is made through the flesh it must be sewed up, as it would permit the stuffing to escape. For the stuffing, the following is extra nice: Peel 4 apples and core them, chop fine with 2 large onions, 4 leaves of sage, and 4 leaves of lemon thyme. Boil some white potatoes, mash them and add 1 pint to the chopped ingredients with a teaspoon of salt and a little cayenne. Stuff the shoulder with this and sew up all the openings. Dredge with flour, salt and pepper and roast in a hot oven, allowing twenty minutes to the pound. Baste frequently, with hot water at first, and then with gravy from the pan. Serve with currant jelly, potatoes and some green vegetables. Another extra good stuffing for pork is made with sweet potatoes as a basis. Boil the potatoes, peel and mash. To a half pint of potato add a quarter pint of finely chopped celery, 2 tablespoons chopped onions, 1/2 teaspoon pepper, teaspoon each of salt and chopped parsley and a tablespoon of butter.

PORK ROASTED WITH TOMATOES.

Take a piece for roasting and rub well with salt and pepper, dredge with flour, and pour into the pan a pint of hot water, and place in a brisk oven. This must be done two or three hours before dinner, according to the size of roast; baste the meat often. An hour before dinner peel some tomatoes (about a quart), put them into a bowl and mash with the hands till the pulp is in fine pieces, add to them a chopped onion, a teaspoon of chopped parsley and 1/2 teaspoon each of sage and thyme. Draw the pan containing the roast to the mouth of oven and skim all the fat from the gravy; pour the tomatoes into the pan, and bake for one hour. With this serve a big dish of rice.

PORK WITH SWEET POTATOES.

Prepare the roast as described above, either stuffed or otherwise. When partly done, peel and cut some sweet potatoes into slices about three inches long. Bank these all around the meat, covering it and filling the pan. Baste often with the gravy and bake one hour. Serve with this a Russian salad, made of vegetables. Young carrots may be used in place of sweet potatoes.

RARE OLD FAMILY DISHES, DESCRIBED FOR THIS WORK BY THE BEST COOKS IN AMERICA. EVERY ONE OF THESE RECIPES IS A SPECIAL FAVORITE THAT HAS BEEN OFTEN TRIED AND NEVER FOUND WANTING. NONE OF THESE RECIPES HAS EVER BEFORE BEEN PRINTED, AND ALL WILL BE FOUND SIMPLE,

ECONOMICAL AND HYGIENIC.

Ham.

BOILED.

Wash well a salted, smoked pig's ham, put this in a large kettle of boiling water and boil until tender, remove from the kettle, take off all of the rind, stick in a quantity of whole cloves, place in a baking pan, sprinkle over with a little sugar, pour over it a cup of cider, or, still better, sherry. Place in the oven and bake brown.

FOR LUNCH.

Mince cold ham fine, either boiled or fried, add a couple of hard-boiled eggs chopped fine, a tablespoon of prepared mustard, a little vinegar and a sprinkling of salt. Put in a mold. When cold cut in thin slices or spread on bread for sandwiches.

BONED.

Having soaked a well-cured ham in tepid water over night, boil it until perfectly tender, putting it on in warm water; take up, let cool, remove the bone carefully, press the ham again into shape, return to the boiling liquor, remove the pot from the fire and let the ham remain in it till cold. Cut across and serve cold.

POTTED.

Mince left-over bits of boiled ham and to every 2 lbs. lean meat allow 1/2 lb. fat. Pound all in a mortar until it is a fine paste, gradually adding 1/2 teaspoon powdered mace, the same quantity of cayenne, a pinch of allspice and nutmeg. Mix very thoroughly, press into tiny jars, filling them to within an inch of the top; fill up with clarified butter or drippings and keep in a cool place. This is nice for tea or to spread picnic sandwiches.

STEW.

A nice way to use the meat left on a ham bone after the frying slices are removed is to cut it off in small pieces, put into cold water to cover and let it come to a boil. Pour off the water and add enough hot to make sufficient stew for your family. Slice an onion and potatoes into it.

WITH VEAL.

A delicious picnic dish is made of ham and veal. Chop fine equal quantities of each and put into a baking dish in layers with slices of hard-boiled eggs between; boil down the water in which the veal was cooked, with the bones, till it will jelly when cold; flavor with celery, pepper and salt and pour over the meat. Cover with a piecrust half an inch thick and bake until the crust is done. Slice thin when cold.

OMELET.

Beat 6 eggs very light, add 1/2 teaspoon salt, 3 tablespoons sweet milk, pepper to taste, have frying pan very hot with 1 tablespoon butter in; turn in the mixture, shake constantly until cooked, then put 1 cup finely chopped ham over the top and roll up like jelly cake, cut in slices.

BAKED.

Most persons boil ham. It is much better baked, if baked right. Soak it for an hour in clean water and wipe dry. Next spread it all over with thin batter and then put it into a deep dish, with sticks under it to keep it out of the gravy. When it is fully done, take off the skin and batter crusted upon the flesh side, and set away to cool. It should bake from six to eight hours. After removing the skin, sprinkle over with two tablespoonfuls of sugar, some black pepper and powdered crackers. Put in pan and return to the oven to brown; then take up and stick cloves through the fat, and dust with powdered cinnamon.

WITH CORN MEAL.

Take bits of cold boiled ham, cut into fine pieces, put in a frying pan with water to cover, season well. When it boils, thicken with corn meal, stirred in carefully, like mush. Cook a short time, pour in a dish to mold, slice off and fry.

BALLS.

Chop 1/2 pint cold boiled ham fine. Put a gill of milk in a saucepan and set on the fire. Stir in 1/2 teacup stale bread crumbs, the beaten yolks of 2 eggs and the ham. Season with salt, cayenne and a little nutmeg. Stir over the fire until hot, add a tablespoon chopped parsley, mix well and turn out to cool. When cold make into small balls, dip in beaten egg, then in bread crumbs and fry in boiling fat.

TOAST.

Remove the fat from some slices of cold boiled ham, chop fine. Put 2 tablespoonfuls of butter into a saucepan on the stove, add the chopped ham and half a cup of sweet cream or milk. Season with pepper and salt; when hot, remove from the stove and stir in quickly 3 well-beaten eggs. Pour onto toast and serve at once.

FLAVORED WITH VEGETABLES.

Take a small ham, as it will be finer grained than a large one, let soak for a few hours in vinegar and water, put on in hot water, then add 2 heads of celery, 2 turnips, 3 onions and a large bunch of savory herbs. A glass of port or sherry wine will improve the flavor of the ham. Simmer very gently until tender, take it out and remove the skin, or if to be eaten cold, let it remain in the liquor until nearly cold.

PATTIES.

One pint of ham which has previously been cooked, mix with two parts of bread crumbs, wet with milk. Put the batter in gem pans, break 1 egg over each, sprinkle the top thickly with cracker crumbs and bake until brown. A nice breakfast dish.

PATTIES WITH ONIONS.

Two cups bread crumbs moistened with a little milk, and two cups cooked ham thoroughly mixed. If one likes the flavor, add a chopped onion. Bake in gem pans. Either break an egg over each gem or chop cold hard-boiled egg and sprinkle over them. Scatter a few crumbs on top. Add bits of butter and season highly with pepper and salt, and brown carefully.

FRIED PATTIES.

One cup cold boiled ham (chopped fine), 1 cup bread crumbs, 1 egg, salt and pepper to taste, mix to the right thickness with nice meat dressing or sweet milk, mold in small patties and fry in butter.

HAM SANDWICHES.

Mince your ham fine and add plenty of mustard, 3 eggs, 1 tablespoon flour, 1 tablespoon butter and as much chopped cucumber pickles as you have ham. Beat this thoroughly together and pour into 1 pint of boiling vinegar, but do not let the mixture boil. When it cools, spread between your sandwiches.

Salt Pork.

FRIED WITH FLOUR.

Slice the pork thinly and evenly, placing it in a large frying pan of water, and turning it twice while freshening. This prevents it humping in the middle, as pork, unless the slices are perfectly flat, cannot be fried evenly. When freshened sufficiently, drain, throw the water off, and, rolling each slice in flour, return to the frying pan. Fry a delicate brown, place on a platter dry, add slices of lemon here and there. Drain all the frying fat off, leaving a brown sediment in the pan. Pour 1 cup of rich milk on this, and when it thickens (keep stirring constantly until of the consistency of rich, thick cream), pour into a gravy boat, and dust with pepper.--[M. G.

FRIED PORK AND GRAVY.

Cut the rind from a firm piece of fat salt pork that has a few streaks of lean (if preferred). Slice thin, scald in hot water, have the frying pan smoking hot, put in the slices of pork and fry (without scorching) until crisp. Then pour off nearly all the fat, add some hot water after the slices have been removed from the pan, and stir in some flour moistened with cold water for a thickened gravy.--[Farmer's Wife.

FRIED IN BATTER OR WITH APPLES.

Slice thin and fry crisp in a hot frying pan, then dip in a batter made as follows: One egg well beaten, 3 large spoons rich milk, and flour enough to make a thin batter. Fry once more until the batter is a delicate brown, and if any batter remains it may be fried as little cakes and served with the pork. Instead of the batter, apples, sliced, may be fried in the fat, with a little water and sugar added, or poor man's cakes, made by scalding 4 spoons granulated (or other) corn meal with boiling water, to which add a pinch of salt and 1 egg, stirred briskly in.--[F. W.

SWEET FRIED.

Take nice slices of pork, as many as you need, and parboil in buttermilk for five minutes, then fry to a golden brown. Or parboil the slices in skimmilk, and while frying sprinkle on each slice a little white sugar and fry a nice brown. Be watchful while frying, as it burns very easily after the sugar is on.--

[I. M. W.

TO FRY IN BATTER.

Prepare as for plain fried pork, fry without dipping in flour, and when done, dip into a batter made as follows: One egg beaten light, 2 tablespoonfuls of milk and the same of sifted flour, or enough to make a thin batter. Stir smooth, salt slightly, dip the fried pork into it and put back into the hot drippings. Brown slightly on both sides, remove to a hot platter and serve immediately.--[R. W.

FRIED WITH SAGE.

Freshen the pork in the usual manner with water or soaking in milk, partly fry the pork, then put three or four freshly picked sprigs of sage in the frying pan with the pork. When done, lay the crisp fried sage leaves on platter with the pork.--[Mrs. W. L. R.

MRS. BISBEE'S CREAMED PORK.

Slice as many slices as your frying pan will hold, pour on cold water, place upon the range to freshen; when hot, pour off the water and fry until crispy; take out upon a platter, pour the fat in a bowl. Pour some milk, about a pint, in the frying pan, boil, thicken and pour upon the fried pork. Serve at once.--[Mrs. G. A. B.

BAKED.

Take a piece of salt pork as large as needed, score it neatly and soak in milk and water half an hour, or longer if very salt; put into a baking pan with water and a little flour sprinkled over the scoring. Bake until done. Always make a dressing to eat with this, of bread and cracker crumbs, a lump of butter, an egg, salt, pepper and sage to taste; mix with hot milk, pack in a deep dish and bake about twenty minutes. Keep water in the baking dish after the meat is taken up, pour off most of the fat and thicken the liquor. Tomatoes go well with this dish, also cranberry sauce.

BOILED.

Boil 4 or 5 lbs. of pork having streaks of lean in it, in plenty of water, for one and one-half hours. Take out, remove skin, cut gashes across the top, sprinkle over powdered sage, pepper and rolled crackers. Brown in the oven. Slice when cold.

CREAMED IN MILK AND WATER.

Freshen 10 or 12 slices of fat pork and fry a nice brown, then take up the pork and arrange on a deep platter. Next pour off half the fat from the frying pan and add 1 cup of milk and 1 of boiling water, and 1 tablespoon flour mixed with a little cold milk or water, or else sifted in when the milk and water begin to boil, but then a constant stirring is required to prevent it from being lumpy. Next add a pinch of salt and a dust of pepper, let it boil up, and pour over the pork. Enough for six.

EGG PORK.

Take slices of pork and parboil in water, sprinkle a little pepper on the pork and put into the frying pan with a small piece of butter and fry. Take 1 egg and a little milk and beat together. When the meat is nearly done, take each slice and dip into the egg, lay back in the pan and cook until done.

CREAMED PORK.

Take 6 slices nice pork, or as many as will fry in the frying pan, and parboil for five minutes, then take out of the water and roll one side of each slice in flour and fry to a golden brown. When fried, turn nearly all of the fat off and set the pan on the stove again and turn on a cup of nice sweet cream; let it boil up, then serve on a platter.

Soups, Stews, Etc.

PORK SOUP.

Put pork bones in pot of cold salted water. Add the following ingredients, in a cheesecloth bag: A few pepper seeds, a bit of horse-radish, mace, and 1 sliced turnip. Boil as for beef soup; strain and add a teaspoon of rice flour to each pint, and let come to a boil. Serve with crackers.

PORK STEW

Slice and fry in a kettle from 1/4 to 1/2 lb. salt pork, drain off the fat and save for shortening, add 3 pints boiling water, 2 or 3 onions sliced thin, 1 quart potatoes sliced and pared, a sprinkling of pepper, large spoon flour mixed in 1 cup of cold water. Let the onions boil a few moments before adding the potatoes and flour. Five minutes before serving, add 1 dozen crackers, split and moistened with hot water, or make dumplings as for any stew.

DRY STEW.

Place slices of pork in the frying pan and fill full with chipped potatoes; pour over a little water and cover tightly, and cook until the pork begins to fry, then loosen from the bottom with a wide knife and pour over more water, and so on until done. Pepper and salt and a bit of butter.

OLD-FASHIONED STEW.

Place 6 large slices of pork in the kettle with nearly a quart of water, let it boil half an hour, then add 8 sliced potatoes and 2 sliced onions, and when nearly done add a little flour, pepper and salt, and a lump of butter.

CHOWDER.

Cut 4 slices of salt pork in dice, place in kettle and fry, add 6 good-sized onions chopped fine, let fry while preparing 8 potatoes, then add 1 quart boiling water and the potatoes sliced thin. Season with salt and pepper to taste. Boil one-half hour.

Miscellaneous.

BACON, BROILED OR FRIED.

The first essential is to have the bacon with a streak of lean and a streak of fat, and to cut or slice it as thin as possible. Then lay it in a shallow tin and set it inside a hot stove. It will toast evenly and the slices will curl up and be so dry that they may be taken in the fingers to eat. The lard that exudes may be thickened with flour, a cup of sweet new milk and a pinch of black pepper added, and nice gravy made. Or if preferred, the bacon, thinly sliced, may be fried on a hot skillet, just turning it twice, letting it slightly brown on both sides. Too long in the hot skillet, the bacon gets hard and will have a burned taste.

BRAINS.

Lay the brains in salt and water for an hour to draw out the blood. Pick them over and take out any bits of bone and membrane. Cook for half an hour in a small quantity of water. When cooked drain off the water, and to each brain add a little pepper, nearly an even teaspoon of salt, a tablespoon of butter and 1 beaten egg. Cook until the egg thickens. Or when the brains are cooked, drain off the water, season with salt, pepper and sage.

PORK AND BEANS.

Pick over and let soak over night 1 quart beans; in the morning wash and drain, and place in a kettle with cold water, with 1/2 teaspoon soda, boil about twenty minutes, then drain and put in earthen bean dish with 2 tablespoons molasses, season with pepper. In the center of the beans put 1 lb. well-washed salt pork, with the rind scored in slices or squares, rind side uppermost. Cover all with hot water and bake six hours or longer, in a moderate oven. Keep covered so they will not burn on the top, but an hour or so before serving remove the pork to another dish and allow it to brown. Beans should also brown over the top.

BOILED DINNER.

Put a piece of salt pork to cook in cold water about 9 o'clock. At 10 o'clock add a few beets, at 11 o'clock a head of cabbage, quartered. One-half hour later add the potatoes. Serve very hot.

GERMAN WICK-A-WACK.

Save the rinds of salt pork, boil until tender, then chop very fine, add an equal amount of dried bread dipped in hot water and chopped. Season with salt, pepper and summer savory; mix, spread one inch deep in baking dish, cover with sweet milk. Bake one-half hour. Very nice.

BROILED PORK.

Soak the pork in cold water over night. Wipe dry and broil over coals until crisp. Pour over it 1/2 pint sweet cream. Ham cooked this way is delicious.

LUNCH LOAF.

Chop remnants of cold boiled ham or salt pork, add crushed crackers and from 3 to 6 eggs, according to the amount of your meat. Bake in a round baking powder box, and when cold it can be sliced for the table.

PORK HASH.

Take scraps of cold pork and ham, chop very fine, put in frying pan, add a very little water, let cook a few minutes, then add twice this amount of chopped potato. Salt and pepper to taste, fry and serve hot.

FOR SUNDAY LUNCHEON.

Take the trimmings saved from ribs, backbone, jowl, shanks of ham and shoulder, and all the nice bits of meat too small for ordinary use; place in a kettle with sufficient water to barely cover meat, and boil slowly until quite tender. Fit a piece of stout cheesecloth in a flat-bottomed dish and cover with alternate strips of fat and lean meat while hot; sprinkle sparingly with white pepper, add another layer of meat and a few very thin slices of perfectly sound tart apples. Repeat until pork is used, then sew up the ends of the cloth compactly, place between agate platters and subject to considerable pressure over night. Served cold this makes a very appetizing addition to Sunday suppers or luncheon.

PORK CHEESE.

Cut 2 lbs. cold roast pork into small pieces, allowing 1/4 lb. fat to each pound of lean; salt and pepper to taste. Pound in a mortar a dessert spoon minced parsley, 4 leaves of sage, a very small bunch of savory herbs, 2 blades of mace, a little nutmeg, half a teaspoon of minced lemon peel. Mix thoroughly with the meat, put into a mold and pour over it enough well-flavored strong stock to make it very moist. Bake an hour and a half and let it cool in the mold. Serve cold, cut in thin slices and garnished with parsley or cress. This is a cooking school recipe. For ordinary use the powdered spices, which may be obtained at almost any country store, answer every purpose. Use 1/4 teaspoon sage, 1/2 teaspoon each of summer savory and thyme, and a pinch of mace.

PORK FLOUR-GRAVY.

Take the frying pan after pork has been fried in it, put in a piece of butter half as large as an egg, let it get very hot, then put in a spoonful of flour sprinkled over the bottom of the pan. Let this get thoroughly browned, then turn boiling water on it, say about a pint. Now take a tablespoon of flour, heaping, wet it up with a cup of sweet milk and stir into the boiling water, add salt and pepper to taste, and a small piece more butter, cook well and serve.

PORK OMELET.

Cut the slices of pork quite thin, discarding the rind, fry on both sides to a light brown, remove from the spider, have ready a batter made of from 2 or 3 eggs (as the amount of pork may require), beaten up with a little flour and a little sweet milk, pouring half of this batter into the spider. Then lay in the pork again, and pour the remaining part of the batter over the pork. When cooked on the one side, cut in squares and turn. Serve hot. Sometimes the pork is cut in small squares before adding the batter.

ANOTHER OMELET.

Put 1 cup cold fried salt pork (cut in dice) and 3 tablespoons sweet milk on back of stove to simmer, then beat 6 eggs and 1 teaspoon salt until just blended. Put 2 tablespoons butter in frying pan. When hot add eggs and shake vigorously until set, then add the hot creamed pork, spread over top, fold, and serve immediately.

PIG'S FEET.

Cut off the feet at the first joint, then cut the legs into as many pieces as there are joints, wash and scrape them well and put to soak over night in cold, slightly salted water; in the morning scrape again and change the water; repeat at night. The next morning put them on to boil in cold water to cover, skim carefully, boil till very tender, and serve either hot or cold, with a brown sauce made of part of the water in which they were boiled, and flavored with tomato or chopped cucumber pickles. If the pig's feet are cooled and then browned in the oven, they will be much nicer than if served directly from the kettle in which they were boiled. Save all the liquor not used for the sauce, for pig's feet are very rich in jelly; when cold, remove the fat, which should be clarified, and boil the liquor down to a glaze; this may be potted, when it will keep a long time and is useful for glazing, or it may be used for soups either before or after boiling, down.--[R. W.

PICKLED PIG'S FEET.

Clean them well, boil until very tender, remove all the bones. Chop the meat, add it to the water they were boiled in, salt to taste. Add enough vinegar to give a pleasing acid taste, pour into a dish to cool. When firm, cut in slices. Or leave out the vinegar and serve catsup of any kind with the meat. Or before cooking the feet, wrap each one in cloth and boil seven hours. When cold take off the cloth and cut each foot in two pieces. Serve cold with catsup or pepper sauce or horse-radish. Or the feet may be put into a jar and covered with cold vinegar, to which is added a handful of whole cloves.--[A. L. N.

KIDNEY ON TOAST.

Cut a kidney in large pieces and soak in cold water an hour. Drain and chop fine, removing all string and fiber; also chop separately one onion. Put a tablespoonful of butter in a frying pan, and when melted add the chopped kidney and stir till the mixture turns a whitish color, then add the onion. Cook five minutes, turn into a small stewpan, season and add a cupful of boiling water. Simmer an hour and thicken with a teaspoonful of cornstarch wet with cold water. Cook five minutes longer, pour over slices of nicely browned toast and serve.

Pork Fritters.

CORN MEAL FRITTERS.

Make a thick batter of corn meal and flour, cut a few slices of pork and fry until the fat is fried out; cut a few more slices, dip them in the batter, and drop them in the bubbling fat, seasoning with salt and pepper; cook until light brown, and eat while hot.

FRITTERS WITH EGG.

Fry slices of freshened fat pork, browning both sides, then make a batter of 1 egg, 1 cup milk, 1 teaspoon baking powder sifted through enough flour to make a rather stiff batter, and a pinch of salt. Now remove the pork from the frying pan and drop in large spoonfuls of the batter, and in the center of each place a piece of the fried pork, then cover the pork with batter, and when nicely brown, turn and let the other side brown. Currant jelly is nice with them.

FRICATELLE.

Chop raw fresh pork very fine, add a little salt and plenty of pepper, 2 small onions chopped fine, half as much bread as there is meat, soaked until soft, 2 eggs. Mix well together, make into oblong patties and fry like oysters. These are nice for breakfast. If used for supper, serve with sliced lemon.

CROQUETTES.

Raw pork chopped fine, 2 cups, 1 small onion chopped very fine, 1 teaspoon powdered sage, 1 cup bread crumbs rubbed fine, salt and pepper to taste, 2 eggs beaten light. Mix thoroughly, make small flat cakes, roll lightly in flour and fry in hot lard.

Pork Pies, Cakes and Puddings.

PORK PIE.

Cut fresh pork in small inch and half-inch pieces, allowing both fat and lean. Boil until done in slightly salted water. Lay away in an earthen dish over night. In the morning it will be found to be surrounded with a firm meat jelly. Will not soak pie crust. Make a rich baking powder biscuit paste. Roll out thin, make top and bottom crust, fill with the prepared pork. Bake.--[H. M. G.

A HINT FOR PORK PIE.

Every housekeeper knows how to make pork pie, but not every one knows that if the bottom crust is first baked with a handful of rice to prevent bubbling--the rice may be used many times for the same purpose--and the pork partially cooked before the upper crust is added, the pie will be twice as palatable as if baked in the old way. The crust will not be soggy and the meat juices will not lose flavor by evaporation.--[Mrs. O. P.

PORK PIE WITH APPLES.

Line a deep pudding dish with pie crust. Place a layer of tart apples in the dish, sprinkle with sugar and a little nutmeg, then place a layer of thin slices of fat salt pork (not cooked), sprinkle lightly with black pepper. Continue to add apples and pork until the dish is full. Cover with a crust and bake until the apples are cooked, when the pork should be melted. Serve as any pie.--[M. C.

SPARERIB PIE.

Chop the small mussy pieces of meat, put in a pudding or bread tin, add some of the gravy and a little water. Make a biscuit crust, roll half an inch thick and put over the top and bake. A tasty way is to cut the crust into biscuits, place close together on top of the meat and bake. More dainty to serve than the single crust. A cream gravy or some left from the rib is nice with this pie. Any of the lean meat makes a nice pie, made the same as the above.

PORK CAKE WITHOUT LARD.

Over 1 lb. fat salt pork, chopped very fine, pour a pint of boiling water. While it is cooling, sift 9 cups flour with 1 heaping teaspoon soda and 2 of cream tartar, stir in 2 cups sugar and 1 of molasses, 4 eggs, teaspoon of all kinds spice, 2 lbs. raisins, 1 lb. currants and 1/2 lb. citron. Lastly, thoroughly beat in the pork and water and bake slowly. This will keep a long time.

PORK CAKE.

Take 1/2 cup sugar, 1/2 cup strong coffee, 1/2 cup molasses, 1/2 cup chopped salt pork, 1/4 cup lard, 1 cup raisins, stoned and chopped, 2 cups flour, 2 eggs, 1 teaspoon soda, dissolved in coffee, 1 teaspoon cloves, cinnamon and nutmeg.

PORK PUDDING.

This is made somewhat after the style of the famous English beefsteak pudding--differs only in two points. Cut up the pieces of fresh pork and stew in the skillet, in slightly salted water, till soft. Make a rich biscuit dough or plain pie paste. Line a quart basin and fill with the stewed pork. Add pepper, a few chopped potatoes if desired, cover all with the paste pinched tightly over, tie a small cloth tightly over the basin, then place basin in a larger cloth, gather the corners together and tie snugly over top, boil in a kettle for half an hour. Be sure the water is boiling hot before placing the basin in, and keep it boiling, with a tight lid.

Roasts.

FRESH LEG.

Score the leg with sharp knife in half-inch gashes, fill with a filling made of chopped onion, sage, bread crumbs and mixed with the beaten yolks and whites of 2 eggs, salt; stuff knuckle and gashes also. Pepper freely and roast it well. A leg weighing 8 lbs. requires three hours of a steady fire. Drain off fat from roasting tin and make a brown gravy. Serve with tart apple sauce.

WITH BUTTERMILK.

Take a piece of pork that is quite lean, soak over night in buttermilk and boil until about half done, then put it in the baking pan, cut through the rind in slices, sprinkle with pepper and sugar and bake to a golden brown.

DANISH PORK ROAST.

Braise the roast, and between each slit insert a bit of sage--which may be removed before serving; place in a deep stewpan and fill the corners and crevices with prunes that have been previously soaked in water long enough to regain their natural size. Roast in moderate oven, basting as usual, taking care not to break the prunes. When half done, take up the prunes, remove pits, crush and add to a dressing made as follows: Moisten 2 cups bread crumbs--one-third corn bread is preferable to all wheat--season with salt, pepper and a mere hint of onions. Put into a cheesecloth bag--saltbag if at hand--and bake beside the roast for half an hour, taking care to prevent scorching. Serve in slices with the roast.

SPARERIB.

Season well with salt, pepper and a little sage. Put in roasting pan with a little water, bake a nice brown. By cracking the ribs twice, you can roll up and fasten with skewers, or tie up with coarse twine. Put the stuffing inside, same as turkey. After it is done, take meat from pan. If the water is not all cooked out, set on top of stove until none remains. Pour out the grease, leaving about half a cup. Set back to cool so as not to cook the gravy too fast at first. Stir 2 spoons or more of flour into the grease and let brown. Add boiling water to make the required amount of gravy. Before removing from fire, add 1/2 cup sweet cream. Baked or mashed potatoes with cold slaw are in order with sparerib, with currant, cranberry or apple sauce. Very nice cold with fried potatoes or chips for supper.

Liver.

WITH BACON.

Pour salted boiling water over the liver and let it stand a few minutes, drain and slice. Crisp thin slices of bacon in a hot frying pan, lay them neatly around the edge of a platter or deep dish, and set the dish where it will keep hot. Fry the liver in the drippings from the bacon and put it in the middle of the dish. Pour a little boiling water into the frying pan, season to taste with pepper and salt, thicken with browned flour and pour over the liver or serve separately.-- [R. F.

LIVER AND ONIONS.

Use two frying pans. In both have a generous supply of fryings or salted lard. Cut the liver in thin, even slices, and wash in cold water. Wipe each slice dry before placing it in the hot grease; fill the frying pan full, pepper and salt all, cover with lid and set over a brisk fire. Slice the onions and place them in the second frying pan of hot grease, pepper, salt and stir frequently. Turn the liver once, each slice. When done, place on a platter, with the onions heaped over and around.--[H. M. G.

HASHED.

Parboil the liver, chop it fine and put it into a hot frying pan with just enough of the liquor it was boiled in to moisten it so it won't be hard and dry. When hot, season with salt, pepper and butter, and serve with mashed potato. Or you can chop cold boiled potatoes with the liver and make a regular hash of it if preferred.--[R. L.

Heart.

STUFFED.

Take three hearts, remove the ventricles and dividing wall, wash and wipe out dry. Fill with 3 tablespoons chopped ham, 4 tablespoons bread crumbs, a little melted butter, some pepper and salt; beat up an egg and mix the meat, etc., with as much of the egg as is needed to bind it together. Tie each heart in a piece of cloth and boil three hours, or till tender, in salt and water. Remove the cloths carefully, so as to keep the dressing in place, rub them over with butter and sprinkle with a little flour, and brown in a brisk oven. Reduce the liquor and thicken it. Serve with mashed potatoes and apple jelly.

BOILED.

Make a biscuit dough rather stiff, sprinkle a well-cleaned heart over with a little pepper and salt, roll the heart securely in the biscuit dough, wrap all in a clean white cloth and sew or baste together loosely, then put in a kettle of hot water and boil about four hours. Serve hot by removing cloth and slicing.

Sausage.

SAUSAGE WITH DRIED BEEF.

To 10 lbs. meat allow 5 tablespoons salt, 4 of black pepper, 3 of sage, and 1/2 tablespoon cayenne. Some persons prefer to add a little ginger, thinking that it keeps the sausage from rising on the stomach. Mix the spices thoroughly through the meat, which may be put into skins or muslin bags and hung in a cold, dry place, or partly cooked and packed in jars with a covering of lard. Every housekeeper uses fried and baked sausages, but sausage and dried beef is a more uncommon dish. Cut the sausage into small pieces, put it into a stewpan with water to cover, and put on to cook. Slice the dried beef and tear it into small pieces, removing fat and gristle, and put into the stew pan. When done, thicken slightly with flour, season and stir an egg quickly into it. Don't get the gravy too thick and don't beat the egg--it wants to show in little flakes of white and yellow.--[Rosalie Williams.

SAUSAGE ROLLS.

Make a rich pie paste, roll out thin and cut, with a large cooky cutter or a canister lid, large discs of the paste. Take a small cooked sausage, and placing it on the edge of the circle of paste, roll it up and pinch the ends together. Bake in a quick oven and serve hot or cold.

WITH CABBAGE.

Put some pieces of fat and lean pork through the sausage mill; add a finely chopped onion, pepper, salt and a dash of mace. Cut a large, sound head of cabbage in two, scoop out the heart of both halves and fill with sausage meat; tie up the head securely with stout twine, put into salted water sufficient to cover the cabbage, and boil one hour and a half. Drain thoroughly and save the liquid, which should not exceed one cupful in all. Brown a tablespoonful of butter over a hot fire, stir in a teaspoon of browned flour and add the liquid; pour over cabbage and serve hot.

GOOD SAUSAGE.

This sausage recipe has been proved good. Take 30 lbs. pork and 12 oz. salt, 2 oz. pepper, 2 oz. sage. Put sage in a pan and dry in oven, then sift. You can add two ounces of ground mustard if you wish. Add 2 or 3 lbs. sugar, mix all together, salt, pepper, etc., and mix with meat before it is chopped. After it is well mixed, cut to your liking.

Fresh Pork.

CUTLETS.

Cut them from a loin of pork, bone and trim neatly and cut away most of the fat. Broil fifteen minutes on a hot gridiron, turning them three or four times, until they are thoroughly done but not dry. Dish, season with pepper and salt and serve with tomato sauce or with small pickled cucumbers as a garnish.

BREADED CUTLETS.

A more elaborate dish is made by dipping the cutlets into beaten egg seasoned to taste with salt, pepper and sage, then into rolled cracker or bread crumbs. Fry slowly till thoroughly done, and serve with mashed potatoes.

CUTLETS FROM COLD ROAST PORK.

Melt an ounce of butter in a saucepan, lay in the cutlets and an onion chopped fine, and fry a light brown; then add a dessertspoon of flour, half a pint of gravy, pepper and salt to taste, and a teaspoon each of vinegar and made mustard. Simmer gently a few minutes and serve.

PORK CHOPS.

The white meat along the backbone (between the ribs and ham) is not always sufficiently appreciated, and is often peeled from the fat, cut from the bones and put into sausage, which should never be done, as it is the choicest piece in the hog to fry. Leave fat and lean together, saw through the bone, fry or broil. The meat gravy should be served in a gravy boat.

BREADED PORK CHOPS.

Cut chops about an inch thick, beat them flat with a rolling pin, put them in a pan, pour boiling water over them, and set them over the fire for five minutes; then take them up and wipe them dry. Mix a tablespoon of salt and a teaspoon of pepper for each pound of meat; rub each chop over with this, then dip, first into beaten egg, then into crackers, rolled, as much as they will take up. Fry in hot lard.

BARBECUED PORK.

Put a loin of pork in a hot oven without water, sprinkle with flour, pepper and salt, baste with butter, cook two or three hours, or until very brown. Pour in the gravy half a teacup of walnut catsup. Serve with fried apples.

Roast Pig.

SUCKING PIG.

Scald carefully and scrape clean, wipe dry, chop off the toes above first joint, remove entrails, and although some cook head entire, it is not advisable. Remove brains, eyes, upper and lower jaws, leaving skin semblance of head, with ears thoroughly scraped and cleaned. Make a dressing composed of one large boiled onion chopped, powdered sage, salt, pepper, 4 cups stale bread crumbs, a bit of butter, and all mixed with well-beaten eggs. Stuff the body part with this. Stitch it up. Previously boil the heart in salted water and stuff this into the boneless head skin to preserve its shape and semblance. Place it down on its feet, head resting on front feet, hind legs drawn out, just as you want it to lie on the platter when served or sent to table. Roast three hours, constantly basting.

TO ROAST WHOLE.

A pig ought not to be under four nor over six weeks old, and ought to be plump and fat. In the city, the butcher will sell you a shoat already prepared, but in the country, we must prepare our own pig for roasting. As soon as the pig is killed, throw it into a tub of cold water to make it tender; as soon as it is perfectly, cold, take it by the hind leg and plunge into scalding water, and shake it about until the hair can all be removed, by the handful at a time. When the hair has all been removed, rub from the tail up to the end of the nose with a coarse cloth. Take off the hoofs and wash out the inside of the ears and nose until perfectly clean. Hang the pig up, by the hind legs, stretched open so as to take out the entrails; wash well with water with some bicarbonate of soda dissolved in it; rinse again and again and let it hang an hour or more to drip. Wrap it in a coarse, dry cloth, when taken down, and lay in a cold cellar, or on ice, as it is better not to cook the pig the same day it is killed. Say kill and clean it late in the evening and roast it the next morning. Prepare the stuffing of the liver, heart and haslets, stewed, seasoned and chopped fine. Mix with these an equal quantity of boiled Irish potatoes, mashed, or bread crumbs, and season with hard-boiled eggs, chopped fine, parsley and sage, or thyme, chopped fine, pepper and salt. Scald the pig on the inside, dry it and rub with pepper and salt, fill with the stuffing and sew up. Bend the forelegs under the body, the hind legs forward, and skewer to keep in position. Place in a large baking pan and pour over it one quart of boiling water. Rub fresh butter all over the pig and sprinkle pepper and salt over it, and put a bunch of parsley and thyme, or sage, in the water. Turn a pan down over it and let it simmer in a hot oven till perfectly tender. Then take off the pan that covers the pig, rub it with more butter and let brown, basting it frequently with the hot gravy. If the hot water and gravy cook down too much, add more hot water and baste. When of a fine brown, and tender and done all through, cover the edges of a large, flat china dish with fresh green parsley and place the pig, kneeling, in the center of the dish. Place in its mouth a red apple, or an ear of green corn, and serve hot with the gravy; or serve cold with grated horse-radish and pickle. Roast pig ought to be evenly cooked, through and through, as underdone pork of any kind, size or age is exceedingly unwholesome. It ought also to be evenly and nicely browned on the outside, as the tender skin when cooked is crisp and palatable. It is easily scorched, therefore keep a pig, while roasting, covered till tender and almost done.

Tongue.

The tongues should be put into the pickle with the hams; boil after three or four weeks, pickle in vinegar which has been sweetened. Add a tablespoon ground mustard to a pint of vinegar. Will keep months. They should be pickled whole. Also nice when first cooked without pickling. Slice cold, to be eaten with or without mayonnaise dressing. Sliced thin, and placed between thin slices of bread, make delicious sandwiches. Chopped fine, with hard-boiled eggs and mayonnaise, make nice sandwiches. Many boil pork and beef tongues fresh. An old brown tongue is an abomination. The saltpeter gives the pink look canned tongues have; the salt and sugar flavor nicely.

When fresh, tongues are nice for mince pies. They may be corned with the hams and boiled and skinned and hot vinegar seasoned with salt and pepper poured over them; or are nice sliced with cold potatoes, garnished with cress or lettuce and a cream salad dressing poured over them. Cream salad dressing: Stir thoroughly together 1 teaspoon sugar, six tablespoons thick sweet cream and 2 tablespoons vinegar, salt and pepper or mustard to taste. The cream and vinegar should be very cold, and the vinegar added to the cream a little at a time, or it will curdle. Stir till smooth and creamy.

Souse.

Take off the horny parts of feet by dipping in hot water and pressing against them with a knife. Singe off hair, let soak in cold water for 24 hours, then pour on boiling water, scrape thoroughly, let stand in salt and water a few hours; before boiling wrap each foot in a clean white bandage, cord securely to keep skin from bursting, which causes the gelatine to escape in the water. Boil four hours. Leave in bandage until cold. If you wish to pickle them, put in a jar, add some of the boiling liquor, add enough vinegar to make a pleasant sour, add a few whole peppers. Very nice cold. If you want it hot, put some of the pickle and feet in frying pan. When boiling, thicken with flour and serve hot.--[Nina Gorton.

See that the feet are perfectly clean, the toes chopped off and every particle cleanly scraped, washed and wiped. Boil for three hours continually, or until every particle falls apart, drain from liquid, pick out all the bones, chop slightly, return to the liquid, add 1/2 cup vinegar, 2 tablespoons sugar, pepper, salt and a dash of nutmeg. (Do not have too much liquid.) Boil up once more and turn all out into a mold, press lightly, and cut cold.--[H. M. Gee.

Thoroughly clean the pig's feet and knock off the horny part with a hatchet. Pour boiling water over them twice and pour it off, then put them on to cook in plenty of water. Do not salt the water. Boil until very tender, then take out the feet, pack in a jar, sprinkle each layer with salt, whole pepper and whole cloves, and cover with equal portions of vinegar and the broth in which the feet were boiled. Put a plate over the top with a weight to keep the souse under the vinegar. If there remains any portion of the broth, strain it and let stand until cold, remove the fat and clarify the broth with a beaten white of egg. It will be then ready for blancmange or lemon jelly and is very delicate.

Scrapple.

Take hog's tongue, heart, liver, all bones and refuse trimmings (some use ears, snout and lights, I do not), soak all bloody pieces and wash them carefully, use also all clean skins, trimmed from lard. Put into a kettle and cover with water, boil until tender and bones drop loose, then cut in sausage cutter while hot, strain liquor in which it was boiled, and thicken with good corn mush meal, boil it well, stirring carefully to prevent scorching. This mush must be well cooked and quite stiff, so that a stick will stand in it. When no raw taste is left, stir in the chopped meat and season to taste with salt, pepper and herb, sage or sweet marjoram, or anything preferred. When the meat is thoroughly mixed all through the mush, and seasoning is satisfactory, dip out into pans of convenient size, to cool. Better lift off fire and stir carefully lest it scorch. When cold, serve in slices like cheese, or fry like mush (crisp both sides) for breakfast, serving it with nice tomato catsup. It tastes very much like fried oysters. Some prefer half buckwheat meal and half corn. To keep it, do not let it freeze, and if not covered with grease melt some lard and pour over, or it will mold. This ought to be sweet and good for a month or more in winter, but will crumble and fry soft if it freezes.--[Mrs. R. E. Griffith.

Head Cheese.

Have the head split down the face, remove the skin, ears, eyes and brains, and cut off the snout; wash thoroughly and soak all day in cold salted water; change the water and soak over night, then put on to cook in cold water to cover. Skim carefully and when done so the bones will slip out, remove to a hot pan, take out every bone and bit of gristle, and chop the meat with a sharp knife as quickly as possible, to keep the fat from settling in it. For 6 lbs.

meat allow 2 tablespoons salt, 1 teaspoon black pepper, a little cayenne, 1/4 teaspoon clove and 2 tablespoons sage. Stir the meat and seasoning well together and put into a perforated mold or tie in a coarse cloth, put a heavy weight on it and let it stand till cold and firm. The broth in which the meat was cooked may be used for pea soup, and the fat, if clarified, may be used for lard.--[R. W.

Cut the head up in suitable pieces to fit the receptacle you wish to boil it in, first cutting off all pieces that are not to be used. If too fat, cut off that, too, and put with the lard to be rendered. Take out the brains and lay them in a dish of cold water, then put the head on to boil till tender. Be sure to skim well. When it begins to boil, cook till the meat is ready to drop off the bones, then take up, remove all bones or gristle and grind or chop, not too fine; put in salt, pepper and cloves to taste, also sage if liked, mix all well together, heat it all together, and pour in a cloth, which is laid in a crock, tie it up tight and put on a weight, to press it. Next day remove the cloth and the head cheese is ready for the table. Skim the fat off the liquor the head was boiled in and set aside for future use. Heat the liquor to a boil and stir in nicely sifted corn meal. After salting, take up in crock and let it get cold, then cut off in slices and fry a nice brown. Nice for breakfast.--[Mrs. A. Joseph.

Pig's Head.

English Brawn: Cut off the hearty cheek or jowl, and try it out for shortening. Saw the pig's head up in small pieces, carefully removing the brains, snoot, eyes, jawbones or portions of teeth sockets. (It is surprising with saw and a keen, sharp-pointed knife how much of the unpleasant pieces of a pig's head can be removed before it is consigned to the salt bath.) Soak all night in salt and water, drain in the morning and set over the fire to boil in slightly salted water. Place the tongue in whole also. When the flesh leaves the bone, take out and strip all into a wooden chopping bowl, reserving the tongue whole. Skin the tongue while warm. Chop the head pieces fine, add pepper, salt, powdered sage to suit taste. Pack all in a deep, narrow mold and press the tongue whole into the middle of the mass. Weight down and set away all night to cool. Keep this always in a cold place until all is used, and, as usual, use a sharp knife to slice.--[Aunt Ban.

To Keep Hams and Shoulders.

We pack them for a few days with a sprinkle of dry salt, then lift and wipe dry (both barrel and meat), repack and cover with brine, which may be prepared thus: To 16 gals. brine (enough to carry an egg) placed in a kettle to boil add 1/4 lb. saltpeter, 3 pts. syrup molasses and a large shovel of hickory ashes tied in a clean saltbag or cloth; boil, skim and cool.--[Mrs. R. E. Griffith.

To prepare smoked ham for summer use: Slice the ham and cut off the rind. Fill a spider nearly full, putting the fat pieces on top. Place in the oven and bake. When partly cooked, pack the slices of hot ham closely in a stone jar and pour the meat juice and fat over the top. Every time that any of the meat is taken out, a little of the lard should be heated and poured back into the jar to keep the meat fresh and good. Be very careful each time to completely cover the meat with lard.--[Marion Chandler.

www.ingramcontent.com/pod-product-compliance
Lightning Source LLC
Chambersburg PA
CBHW070328190526
45169CB00005B/1792